Methods in Enzymology

Volume 367
LIPOSOMES
Part A

METHODS IN ENZYMOLOGY

EDITORS-IN-CHIEF

John N. Abelson   Melvin I. Simon

DIVISION OF BIOLOGY
CALIFORNIA INSTITUTE OF TECHNOLOGY
PASADENA, CALIFORNIA

FOUNDING EDITORS

Sidney P. Colowick and Nathan O. Kaplan

Methods in Enzymology

Volume 367

Liposomes

Part A

EDITED BY

Nejat Düzgünes

DEPARTMENT OF MICROBIOLOGY
UNIVERSITY OF THE PACIFIC SCHOOL OF DENISTRY
SAN FRANCISCO, CALIFORNIA

ELSEVIER
ACADEMIC
PRESS

AMSTERDAM • BOSTON • HEIDELBERG • LONDON
NEW YORK • OXFORD • PARIS • SAN DIEGO
SAN FRANCISCO • SINGAPORE • SYDNEY • TOKYO
Academic Press is an imprint of Elsevier

Academic Press
An Elsevier Imprint.
525 B Street, Suite 1900, San Diego, California 92101-4495, USA
http://www.academicpress.com

Academic Press
84 Theobald's Road, London WC1X 8RR, UK
http://www.academicpress.com

International Standard Book Number: 0-12-182270-2

PRINTED IN THE UNITED STATES OF AMERICA
03 04 05 06 07 08   9 8 7 6 5 4 3 2 1

Table of Contents

Section I. Methods of Liposome Preparation

Section II. Physicochemical Characterization of Lipsomes

Contributors to Volume 367

Article numbers are in parentheses and following the names of contributors.
Affiliations listed are current.

PATRICK AHL (80), *Bio Delivery Sciences International, Inc., UMDNJ-New Jersey Medical School, 185 South Orange Avenue, ADMC4, Newark, New Jersey 07103*

JUHA-MATTI ALAKOSKELA (129), *Institute of Biomedicine, P.O. Box 63, Biomedcum Haartmaninkatu 8, University of Helsinki, Helsinki, FIN 00014, Finland*

MIGLENA ANGELOVA (15), *Institute of Biomedicine, P.O. Box 63, Biomedicum Haartmaninkatu 8, University of Helsinki, Helsinki, FIN 00014, Finland*

KLAUS ARNOLD (253), *Institute for Medical Physics and Biophysics, Faculty of Medicine, University of Leipzig, D-04103 Leipzig, Germany*

JESUS ARROYO (213), *Facultad Farmacia y Bioquímica, Universidad de Buenos Aires,Junin 956 2P, Buenos Aires 11113, Argentina*

ANDREW BACON (70), *School of Pharmacy, Lipoxen Technologies Ltd., University of London, 29-39 Brunswick Square, London WC1N 1AX, England*

LUIS A. BAGATOLLI (233), *MEMPHYS-Center for Biomembrane Physics, Department of Biochemistry and Molecular Biology, Campusvej 55, DK-5230 Odense M, Denmark*

YECHEZKEL BARENHOLZ (270), *Laboratory of Membrane and Liposome Research, Hebrew Univeristy-Hadassah Medical School, Jerusalem 91120, Israel*

DELIA L. BERNIK (213), *Facultad Farmacia y Bioquímica, Universidad de Buenos Aires, Junin 956 2P, Buenos Aires 11113, Argentina*

WILSON CAPPARÓS-WANDERLEY (70), *School of Pharmacy, Lipoxen Technologies Ltd., University of London, 29-39 Brunswick Square, London WC1N 1AX, England*

LAURIE CHOW (3), *Inex Pharmaceutical Copre, Glenlyon Business Park1, 100-8900 Glenlyon Parkway, Burnaby, British Columbia, Canada V5J 5J8*

JOEL A. COHEN (148), *Department of Physiology, University of the Pacific School of Dentistry, 2155 Webster Street, San Francisco, California 94115*

RIVKA COHEN (270), *Laboratory of Membrane and Liposome Research, Hebrew Univeristy-Hadassah Medical School, Jerusalem 91120, Israel*

E. ANIBAL DISALVO (213), *Facultad Farmacia y Bioquímica, Universidad de Buenos Aires, Junin 956 2P, Buenos Aires 11113, Argentina*

NEJAT DÜZGÜNES (23), *Department of Microbiology, University of the Pacific School of Dentistry, 2155 Webster Street, San Francisco, California 94115*

SIMCHA EVEN-CHEN (270), *Laboratory of Membrane and Liposome Research, Hebrew Univeristy-Hadassah Medical School, Jerusalem 91120, Israel*

GREGORY GREGORIADIAS (70), *School of Pharmacy, Lipoxen Technologies Ltd., University of London, 29-39 Brunswick Square, London WC1N 1AX, England*

SADAO HIROTA (177), *Tokyo Denki University, 6-6-18 Higashikaigan-Minami, Chigasaki-Shi 253-0054, Japan*

JUHA M. HOLOPAINEN (15), *Institute of Biomedicine, P.O. Box 63, Biomedicum Haartmaninkatu 8, University of Helsinki, Helsinki, FIN 00014, Finland*

REUMA HONEN (270), *Laboratory of Membrane and Liposome Research, Hebrew Univeristy-Hadassah Medical School, Jerusalem 91120, Israel*

MICHAEL HOPE (3), *Inex Pharmaceutical Copre, Glenlyon Business Park1, 100-8900 Glenlyon Parkway, Burnaby, British Columbia, Canada V5J 5J8*

JANA JASS (199), *The Lawson Health Research Institute, 268 Grosvenor Street, London, Ontario, Canada N6A 4U2*

ANDREA KAŠNÁ (111), *Veterinary Research Institute, Department of Immunology, Hudcova 70, 62132 Brno, Czech Republic*

PAAVO K. J. KINNUNEN (15, 129), *Institute of Biomedicine, P.O. Box 63, Biomedicum Haartmaninkatu 8, University of Helsinki, Helsinki, FIN 00014, Finland*

PETER LAGGNER (129), *Institute of Biophysics and X-Ray Structure Research, Austrian Academy of Sciences, Schmiedl-strasse 6, A-8042 Graz, Austria*

BRENDA MCCORMACK (70), *School of Pharmacy, Lipoxen Technologies Ltd., University of London, 29-39 Brunswick Square, London WC1N 1AX, England*

BARBARA MUI (3), *Inex Pharmaceutical Copre, Glenlyon Business Park1, 100-8900 Glenlyon Parkway, Burnaby, British Columbia, Canada V5J 5J8*

JIŘÍ NEČA (111), *Veterinary Research Institute, Department of Immunology, Hudcova 70, 62132 Brno, Czech Republic*

SHINPEI OHKI (253), *Department of Physiology and Biophysics, School of Medicine and Biomedical Sciences, State University of New York at Buffalo, Buffalo, New York 14214*

WALTER R. PERKINS (80), *Transave, Inc., 11 Deerpark Drive, Suite 117, Monmouth Junction, New Jersey 08552*

GERTRUD PUU (199), *Swedish Defense Research Agency, NBC Defence, SE 90182 Umei, Sweden*

RAMON BARNADAS I RODRÍGUEZ (28), *Unitat de Biofisica, Facultat de Medicine, Universitat Autònoma de Barcelona, Catalonia, 08193 Cerdanolya del Vallès, Spain*

ROLF SCHUBERT (46), *Pharmazeutisches Institut, Lehrstuhl für Pharmazeutische Technologie, Albert-Ludwigs-Universität-Freiburg, Hermann-herder Strasse 9, D-79104 Freiburg, Germany*

HILARY SHMEEDA (270), *Shaare Zedek Medical Center, Department of Experimental Oncology, POB 3235, Jerusalem 91031, Israel*

TORBJÖRN TJÄRNHAGE (199), *Swedish Defense Research Agency, NBC Defence, SE 90182 Umei, Sweden*

JAROSLAV TURÁNEK (111), *Veterinary Research Institute, Department of Immunology, Hudcova 70, 62132 Brno, Czech Republic*

CARMELA WEINTRAUB (270), *Laboratory of Membrane and Liposome Research,- Hebrew Univeristy-Hadassah Medical-School, Jerusalem 91120, Israel*

EWOUD C. A. VAN WINDEN (99), *Regulon Gene Pharmaceuticals A.E.B.E., Auxentiou Grigoriou 7, Alimos, 17455 Athens, Greece*

MANUEL SABÉS I XAMANÍ (28), *Unitat de Biofisica, Facultat de Medicine, Universitat Autònoma de Barcelona, Catalonia, 08193 Cerdanolya del Vallès, Spain*

DANA ZÁLUSKÁ (111), *Veterinary Research Institute, Department of Immunology, Hudcova 70, 62132 Brno, Czech Republic*

Preface

The origins of liposome research can be traced to the contributions of Alec Bangham and colleagues in the mid 1960s. The description of lecithin dispersions as containing "spherulites composed of concentric lamellae" (A. D. Bangham and R. W. Horne, J. Mol. Biol. 8, 660, 1964) was followed by the observation that "the diffusion of univalent cations and anions out of spontaneously formed liquid crystals of lecithin is remarkably similar to the diffusion of such ions across biological membranes (A. D. Bangham, M. M. Standish and J. C. Watkins, J. Mol. Biol. 13, 238, 1965). Following early studies on the biophysical characterization of multilamellar and unilamellar liposomes, investigators began to utilize liposomes as a well-defined model to understand the structure and function of biological membranes. It was also recognized by pioneers, including Gregory Gregoriadis and Demetrios Papahadjopoulos, that liposomes could be used as drug delivery vehicles. It is gratifying that their efforts and the work of those inspired by them have led to the development of liposomal formulations of doxorubicin, daunorubicin, and amphotericin B, now utilized in the clinic. Other medical applications of liposomes include their use as vaccine adjuvants and gene delivery vehicles, which are being explored in the laboratory as well as in clinical trials. The field has progressed enormously since 1965.

This volume describes methods of liposome preparation, and the physicochemical characterization of liposomes. I hope that these chapters will facilitate the work of graduate students, post-doctoral fellows, and established scientists entering liposome research. Subsequent volumes in this series will cover additional subdisciplines in liposomology.

The areas represented in this volume are by no means exhaustive. I have tried to identify the experts in each area of liposome research, particularly those who have contributed to the field over some time. It is unfortunate that I was unable to convince some prominent investigators to contribute to the volume. Some invited contributors were not able to prepare their chapters, despite generous extensions of time. In some cases I may have inadvertently overlooked some experts in a particular area, and to these individuals I extend my apologies. Their primary contributions to the field will, nevertheless, not go unnoticed, in the citations in these volumes and in the hearts and minds of the many investigators in liposome research.

In the last five years, the liposome field has lost some of its major members. Demetrios Papahadjopoulos (one of Alec Bangham's proteges and one of my mentors) was a significant mover of the field and an inspiration to many young scientists. He organized the first conference on liposomes in 1977 in New York. He was also a co-founder of a company to attempt to commercialize liposomes for medical purposes. Danilo Lasic brought in his sophisticated biophysics background to help understand liposome behavior, wrote and co-edited numerous volumes on various aspects of liposomes, and helped their widespread appreciation with short reviews. David O'Brien was a pioneer in the field of photoactivatable liposomes, most likely inspired by his earlier work on rhodopsin. He was to have contributed a chapter to the last volume of "Liposomes" in this series. For all their contributions to the field, this volume is dedicated to the memories of Drs. Papahadjopoulos, Lasic and O'Brien.

I would like to express my gratitude to all the colleagues who graciously contributed to these volumes. I would like to thank Shirley Light of Academic Press for her encouragement for this project, and Noelle Gracy of Elsevier Science for her help at the later stages of the project. I am especially thankful to my wife Diana Flasher for her understanding, support and love during the endless editing process, and my children Avery and Maxine for their unique curiosity, creativity, cheer, and love.

NEJAT DÜZGÜNES

MILL VALLEY

METHODS IN ENZYMOLOGY

VOLUME 90. Carbohydrate Metabolism (Part E)
Edited by WILLIS A. WOOD

VOLUME 91. Enzyme Structure (Part I)
Edited by C. H. W. HIRS AND SERGE N. TIMASHEFF

VOLUME 92. Immunochemical Techniques (Part E: Monoclonal Antibodies and
General Immunoassay Methods)
Edited by JOHN J. LANGONE AND HELEN VAN VUNAKIS

VOLUME 93. Immunochemical Techniques (Part F: Conventional Antibodies,
Fc Receptors, and Cytotoxicity)
Edited by JOHN J. LANGONE AND HELEN VAN VUNAKIS

VOLUME 94. Polyamines
Edited by HERBERT TABOR AND CELIA WHITE TABOR

VOLUME 95. Cumulative Subject Index Volumes 61–74, 76–80
Edited by EDWARD A. DENNIS AND MARTHA G. DENNIS

VOLUME 96. Biomembranes [Part J: Membrane Biogenesis: Assembly and
Targeting (General Methods; Eukaryotes)]
Edited by SIDNEY FLEISCHER AND BECCA FLEISCHER

VOLUME 97. Biomembranes [Part K: Membrane Biogenesis: Assembly and
Targeting (Prokaryotes, Mitochondria, and Chloroplasts)]
Edited by SIDNEY FLEISCHER AND BECCA FLEISCHER

VOLUME 98. Biomembranes (Part L: Membrane Biogenesis: Processing and
Recycling)
Edited by SIDNEY FLEISCHER AND BECCA FLEISCHER

VOLUME 99. Hormone Action (Part F: Protein Kinases)
Edited by JACKIE D. CORBIN AND JOEL G. HARDMAN

VOLUME 100. Recombinant DNA (Part B)
Edited by RAY WU, LAWRENCE GROSSMAN, AND KIVIE MOLDAVE

VOLUME 101. Recombinant DNA (Part C)
Edited by RAY WU, LAWRENCE GROSSMAN, AND KIVIE MOLDAVE

VOLUME 102. Hormone Action (Part G: Calmodulin and Calcium-Binding
Proteins)
Edited by ANTHONY R. MEANS AND BERT W. O'MALLEY

VOLUME 103. Hormone Action (Part H: Neuroendocrine Peptides)
Edited by P. MICHAEL CONN

VOLUME 104. Enzyme Purification and Related Techniques (Part C)
Edited by WILLIAM B. JAKOBY

VOLUME 105. Oxygen Radicals in Biological Systems
Edited by LESTER PACKER

VOLUME 106. Posttranslational Modifications (Part A)
Edited by FINN WOLD AND KIVIE MOLDAVE

VOLUME 193. Mass Spectrometry
Edited by JAMES A. MCCLOSKEY

VOLUME 194. Guide to Yeast Genetics and Molecular Biology
Edited by CHRISTINE GUTHRIE AND GERALD R. FINK

VOLUME 195. Adenylyl Cyclase, G Proteins, and Guanylyl Cyclase
Edited by ROGER A. JOHNSON AND JACKIE D. CORBIN

VOLUME 196. Molecular Motors and the Cytoskeleton
Edited by RICHARD B. VALLEE

VOLUME 197. Phospholipases
Edited by EDWARD A. DENNIS

VOLUME 198. Peptide Growth Factors (Part C)
Edited by DAVID BARNES, J. P. MATHER, AND GORDON H. SATO

VOLUME 199. Cumulative Subject Index Volumes 168–174, 176–194

VOLUME 200. Protein Phosphorylation (Part A: Protein Kinases: Assays, Purification, Antibodies, Functional Analysis, Cloning, and Expression)
Edited by TONY HUNTER AND BARTHOLOMEW M. SEFTON

VOLUME 201. Protein Phosphorylation (Part B: Analysis of Protein Phosphorylation, Protein Kinase Inhibitors, and Protein Phosphatases)
Edited by TONY HUNTER AND BARTHOLOMEW M. SEFTON

VOLUME 202. Molecular Design and Modeling: Concepts and Applications (Part A: Proteins, Peptides, and Enzymes)
Edited by JOHN J. LANGONE

VOLUME 203. Molecular Design and Modeling: Concepts and Applications (Part B: Antibodies and Antigens, Nucleic Acids, Polysaccharides, and Drugs)
Edited by JOHN J. LANGONE

VOLUME 204. Bacterial Genetic Systems
Edited by JEFFREY H. MILLER

VOLUME 205. Metallobiochemistry (Part B: Metallothionein and Related Molecules)
Edited by JAMES F. RIORDAN AND BERT L. VALLEE

VOLUME 206. Cytochrome P450
Edited by MICHAEL R. WATERMAN AND ERIC F. JOHNSON

VOLUME 207. Ion Channels
Edited by BERNARDO RUDY AND LINDA E. IVERSON

VOLUME 208. Protein–DNA Interactions
Edited by ROBERT T. SAUER

VOLUME 209. Phospholipid Biosynthesis
Edited by EDWARD A. DENNIS AND DENNIS E. VANCE

Section I

Methods of Liposome Preparation

[1]   Extrusion Technique to Generate Liposomes of Defined Size

By BARBARA MUI, LAURIE CHOW and MICHAEL J. HOPE

Introduction

Liposome extrusion is a widely used process in which liposomes are forced under pressure through filters with defined pore sizes to generate a homogeneous population of smaller vesicles with a mean diameter that reflects that of the filter pore.[1] This technique has grown in popularity and has become the most common method of reducing multilamellar liposomes, usually called multilamellar vesicles (MLVs), to large unilamellar vesicles (LUVs) for model membrane and drug delivery research.

The extrusion concept was initially introduced by Olson *et al.*,[2] who described the sequential passage of a dilute liposome preparation through polycarbonate filters of decreasing pore size, using a hand-held syringe and filter holder attachment, in order to produce a homogeneous size distribution. This procedure was further developed and made more practical by the construction of a robust, metal extrusion device that employed medium pressures (800 lb/in^2) to rapidly extrude MLV suspensions directly through polycarbonate filters with pore diameters in the range of 50 to 200 nm to generate LUVs.[1] At the time this process represented a major advance for those routinely preparing LUVs. Other size reduction methods, such as the use of ultrasound or microfluidization techniques, tend to generate significant populations of "limit size" vesicles that are subject to lipid-packing constraints[3] and also suffer from lipid degradation, heavy metal contamination, and limited trapping efficiencies. Reversed phase evaporation (REV) methods were also common in the 1980s and usually involved the formation of aqueous–organic emulsions followed by solvent evaporation to produce liposome populations with large trapped volumes and improved trapping efficiencies.[4] However, these methods are restricted by lipid solubility in solvent or solvent mixtures; moreover,

---

[1] M. J. Hope, M. B. Bally, G. Webb, and P. R. Cullis, *Biochim. Biophys. Acta* **812,** 55 (1985).

[2] F. Olson, C. A. Hunt, F. C. Szoka, W. J. Vail, and D. Papahadjopoulos, *Biochim. Biophys. Acta* **557,** 9 (1979).

[3] M. J. Hope, M. B. Bally, L. D. Mayer, A. S. Janoff, and P. R. Cullis, *Chem. Phys. Lipids* **40,** 89 (1986).

[4] F. Szoka, F. Olson, T. Heath, W. Vail, E. Mayhew, and D. Papahadjopoulos, *Biochim. Biophys. Acta* **601,** 559 (1980).

removal of residual solvent can be tedious. Detergent dialysis techniques are also subject to similar practical difficulties associated with lipid solubility and complete removal of detergent.

Consequently, the convenience and speed of extrusion became a major advantage over other techniques. Extrusion can be applied to a wide variety of lipid species and mixtures, it works directly from MLVs without the need for sequential size reduction, process times are on the order of minutes, and it is only marginally limited by lipid concentration compared with other methods. Manufacturing issues related to removal of organic solvents or detergents from final preparations are eliminated and the equipment available for extrusion scales well from bench volumes (0.1 to 10 mL) through preclinical (10 mL to 1 liter) to clinical (>1 liter) volumes employing relatively low-cost equipment, especially at the research and preclinical levels.

Extrusion and Extrusion Devices

MLVs form spontaneously when bilayer-forming lipid mixtures are hydrated in excess water, but they exhibit a broad size distribution ranging from 0.5 to 10 μm in diameter and the degree of lamellarity varies depending on the method of hydration and lipid composition. These factors restrict severely the practical application of MLVs for membrane and drug delivery research, as discussed in detail elsewhere.[3] In general, <10% of the total lipid present in a normal multilamellar liposome is present in the outer monolayer of the externally exposed bilayer compared with 50% in the outer monolayer of a large unilamellar system.[1] Consequently, the LUV better reflects the bilayer structure of a typical plasma or large organelle membrane. Other limitations of MLVs include their large diameter, size heterogeneity, multiple internal compartments, low trap volumes, and inconsistencies from preparation to preparation. Therefore, sizing MLV preparations by extrusion is an effective way to overcome some of these problems and to generate reproducible model membrane systems for basic research, applied research, and clinical applications.

Only moderate pressures (typically 200–800 lb/in^2) are required to force liquid crystalline MLVs through polycarbonate filters with defined pore sizes. The majority of laboratories specializing in liposome research, particularly as applied to drug delivery, use a heavy-duty device commercially available from Northern Lipids (Vancouver, BC, Canada; www.northernlipids.com). The Lipex extruder is an easy-to-use, robust stainless steel unit, which can operate up to pressures of 800 lb/in^2 (Fig. 1). A quick-fit sample port assembly allows for rapid and convenient cycling of preparations through the filter holder. The sequential use of large to small pore filters[2] to reduce back pressure is not necessary for

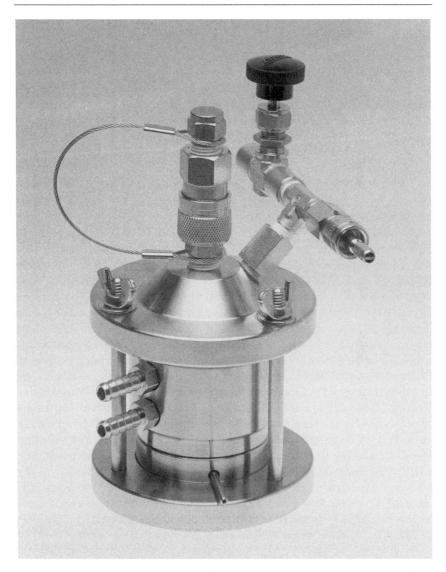

FIG. 1. A research-scale extrusion device (Lipex extruder) manufactured by Northern Lipids (Vancouver, BC, Canada) has a 10-mL capacity and can be operated over a wide range of temperatures when used in combination with a circulating water bath. The quick-release sample port at the top of the unit allows for rapid cycling of sample through the filters.

the majority of lipid samples, and large multilamellar systems can be extruded directly through filter pore sizes as small as 30 nm. The equipment is also fitted with a water-jacketed, sample-holding barrel that enables the extrusion of lipids with gel–liquid crystalline phase transitions above room temperature, an important feature as gel-state lipids will not extrude (see Effect of Lipid Composition on Extrusion, later).

Extrusion can also be performed with a hand-held syringe fitted with a standard sterilization filter holder or purpose-built hand-held units, such as those supplied by Avanti Polar Lipids (Alabaster, AL; www.avanti-lipids.com) and Avestin (Ottawa, ON, Canada; www.avestin.com). These devices are suitable only for small-volume applications (typically <1 mL); one example consists of two Hamilton syringes connected by a filter holder, allowing for back-and-forth passage of the sample.[5] Using this technique, a dilute suspension of liposomes (composed of liquid crystalline lipid) can be passed through the filters to reduce vesicle size. This method, however, is limited by the back pressure that can be tolerated by the syringe and filter holder, as well as the pressure that can be applied manually. Generally, phospholipid concentrations must be less than 30 mM in order to comfortably extrude liposomes manually.

A variety of filters suitable for reducing the mean diameter of liposome preparations are available from scientific suppliers. The most commonly used are standard polycarbonate filters (with straight-through pores). Other filter materials can be used, but the polycarbonate type has proved to be reliable, inert, durable, and easy to apply to filter supports without damage. Pore density influences extrusion pressure. In our experience there is usually little variation between filters from the same manufacturer. However, on occasion users may notice changes in vesicle diameter prepared when using filters from different batches from the same supplier or when using filters in which the pores are created by different manufacturing processes. Tortuous path type filters do not have well-defined pore diameters like the straight-through type, and back pressure tends to be higher when using these filters for liposome extrusion. However, adequate size reduction can still be achieved.

Mechanism of Extrusion and Vesicle Morphology

As the concentric layers of a typical MLV squeeze into the filter pore under pressure during extrusion, a process of membrane rupture and resealing occurs. The practical consequence of this is that any solute trapped

[5] R. C. MacDonald, R. I. MacDonald, B. P. Menco, K. Takeshita, N. K. Subbarao, and L. R. Hu, *Biochim. Biophys. Acta* **1061,** 297 (1991).

inside an MLV or large liposome before size reduction will leak out during the extrusion cycle. Therefore, when specific solutes are to be encapsulated, extrusion is nearly always performed in the presence of medium containing the desired final solute concentration and external (unencapsulated) solute is removed only when sizing is complete. In a study on the mechanism of liposome size reduction by extrusion, Hunter and Frisken[6] demonstrated that the pressure needed to reduce the particle size of vesicles during passage through a 100-nm pore correlated with the force needed to rupture the lipid membrane and not the force required simply to deform the bilayer. Interestingly, these authors also noted that as flow rate through the filter increased the mean vesicle size decreased. This is attributed to the thickness of the lubricating layer formed by fluid associated with the sides of the pore from which particles are excluded. As the velocity of the fluid increases the thickness of the lubricating layer also increases, effectively reducing the pore diameter experienced by vesicles traversing the membrane.[6,7]

The rupture and resealing process can also give rise to oval or sausage-shaped vesicles, and Mui et al.[8] showed that this shape deformation is dictated largely by osmotic force. As vesicles are squeezed through the pores they elongate and lose internal volume through transient membrane rupture to accommodate the increase in surface area-to-volume ratio associated with the nonspherical morphology. On exiting the pore the membrane wants to adopt a spherical shape, thermodynamically the lowest energy state for the bilayer, but the required increase in trapped volume is opposed by osmotic force. Therefore, in the presence of impermeable or semipermeable solutes (e.g., common buffers and salts) oval or sausage-shaped vesicles are produced, whereas vesicles made in pure water are spherical (Fig. 2A and B).

Sausage-like and dimpled vesicle morphology is observed when extrusion occurs even in solutions of relatively low osmolarity, such as 10 mM NaCl. It should be noted that these vesicle morphologies have been observed only when employing cryoelectron microscopy techniques, in which vesicles are visualized through thin films of ice in the absence of cryoprotectants. Freeze–fracture methods do not reveal sausage-like morphology under the same conditions, which may be due to the high concentrations of membrane-permeable glycerol (25%, v/v), used as a cryoprotectant, affecting the osmotic gradient. Rounding up of vesicles is readily achieved by simply lowering the ionic strength of the external medium.[8]

[6] D. G. Hunter and B. J. Frisken, *Biophys. J.* **74,** 2996 (1998).
[7] G. Gompper and D. M. Kroll, *Phys. Rev. E Stat. Phys. Plasmas Fluids Relat. Interdiscip. Topics* **52,** 4198 (1995).
[8] B. L. Mui, P. R. Cullis, E. A. Evans, and T. D. Madden, *Biophys. J.* **64,** 443 (1993).

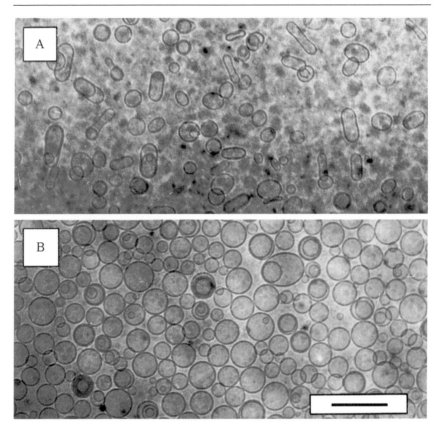

FIG. 2. Cryoelectron microscopy of extruded vesicles. Shown are vesicles of egg phosphatidylcholine–cholesterol (55:45 molar ratio) made in (A) 150 mM NaCl, 20 mM HEPES, pH 7.4, or (B) distilled water. Scale bar: 200 nm.

Formation of Unilamellar Vesicles

The well-defined aqueous compartment and single bilayer free of lipid-packing constraints make LUVs important model systems in membrane and liposomal drug delivery research. Cycling an MLV preparation through filters with 100-nm pores produces a homogeneous population of vesicles with a mean diameter of approximately 100 nm, usually after about 10 passes (Fig. 3). Lamellarity of a liposome preparation can be determined by using ^{31}P nuclear magnetic resonance (NMR) to monitor the phospholipid phosphorus signal intensity at the outer monolayer compared with the total signal. Adding an impermeable paramagnetic or broadening reagent to the external medium will decrease the intensity of

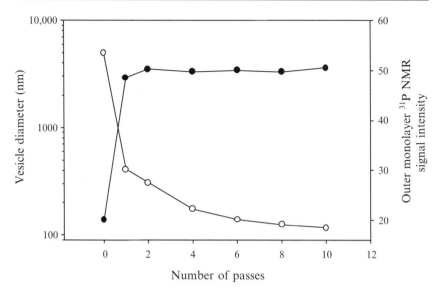

FIG. 3. Vesicle size reduction and increased phospholipid present in the outer monolayer. MLVs (100 mg of egg phosphatidylcholine per milliliter) prepared in 150 mM NaCl, 20 mM HEPES, pH 7.4, were extruded through two stacked 100-nm pore-sized filters. (○) Mean vesicle diameter determined by quasi-elastic light scattering. (●) The percentage of total phospholipid present in the outer monolayer determined by ^{31}P NMR.

the initial ^{31}P NMR signal by an amount proportional to the fraction of lipid exposed to the external medium.[1,9,10] During the first five passes through two (stacked) polycarbonate filters with 100-nm pore sizes, egg phosphatidylcholine (egg PC) MLVs rapidly decrease in size, whereas a concomitant increase in phospholipid detectable at the interface with the external medium is observed (Fig. 3). These data are consistent with the large multilamellar structure, in which the majority of the lipid is associated with internal bilayers, rupturing and resealing into progressively smaller vesicles with fewer and fewer internal lamellae, until approximately 50% of the phosphorous signal is accounted for in the outer monolayer, indicating that the vesicle population largely consists of single bilayer vesicles (LUVs). Between 5 and 10 cycles there is no further change in either mean size or outer monolayer signal intensity.

[9] N. Düzgüneş, J. Wilschut, K. Hong, R. Fraley, C. Perry, D. S. Friend, T. L. James, and D. Papahadjopoulos, *Biochim. Biophys. Acta* **732,** 289 (1983).
[10] L. D. Mayer, M. J. Hope, and P. R. Cullis, *Biochim. Biophys. Acta* **858,** 161 (1986).

A common practice, introduced by Mayer et al.,[11] is to subject MLVs to freeze–thaw cycles before extrusion, which increases the proportion of unilamellar vesicles in preparations sized through filters with a pore size >100 nm. It is important to note, however, that the thawing must occur at temperatures above the gel–liquid crystalline phase transition of the lipids used, unless cholesterol is included in the lipid mixture.[12] It is estimated that as much as 90% of vesicles passed through a filter with a pore size of 200 nm are unilamellar if prepared from frozen and thawed multilamellar vesicles.[10] The freezing and thawing cycle has been shown to cause internal lamellae of MLVs to separate and vesiculate, which probably reduces the number of closely associated bilayers forced through pores together, thus reducing the formation of oligolamellar vesicles. Freeze–fracture electron microscopy gives a more qualitative indication of lamellarity than ^{31}P NMR signal intensity measurements. This technique provides a unique view of internal lamellae when cross-fracturing occurs. Figure 4A is a freeze–fracture micrograph of an egg PC multilamellar liposome that has cross-fractured, thus demonstrating the close apposition and large number of internal bilayers associated with a typical MLV. Figure 4B shows vesicles that have been sized through a 400-nm pore-size filter; some cross-fracturing is visible, revealing the oligolamellar nature of this preparation. However, MLVs extruded through 100-nm–diameter pores consist of single-bilayer vesicles (Fig. 4C). Another key advantage of the extrusion technique is the ability to process liposomes at high lipid concentrations. Figure 5 is a freeze–fracture micrograph of egg PC vesicles sized to 100 nm at a concentration of 400 mg/ml.

Effect of Lipid Composition on Extrusion

Perhaps the most important compositional factor in liposome extrusion is the gel–liquid crystalline phase transition temperature (T_c) of the membrane lipid. Nayar et al.[12] conducted an extensive study on temperature and extrusion of MLVs composed of distearoyl phosphatidylcholine (DSPC) and DSPC–cholesterol (55:45 molar ratio). At an applied pressure of 500 lb/in^2 the flow rate of liposomes through a 25-mm Whatman Nuclepore (Newton, MA) filter with a pore size of 100 nm was recorded as a function of temperature. MLVs composed of DSPC alone could be extruded only above 55° (the T_c for DSPC); at lower temperatures pressures as high as 800 lb/in^2 did not result in extrusion. However, above 55° extrusion

[11] L. D. Mayer, M. J. Hope, P. R. Cullis, and A. S. Janoff, *Biochim. Biophys. Acta* **817,** 193 (1985).
[12] R. Nayar, M. J. Hope, and P. R. Cullis, *Biochim. Biophys. Acta* **986,** 200 (1989).

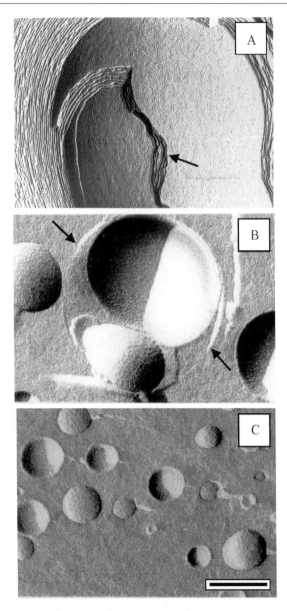

FIG. 4. Vesicle lamellarity visualized by freeze–fracture microscopy. (A) The close apposition and multiple internal bilayers are seen in cross-fracture (arrow) of MLVs prepared from egg PC. (B) Egg PC MLVs extruded through 400-nm pore-size filters producing oligolamellar vesicles (arrows) and (C) single-bilayer vesicles obtained by extrusion through 100-nm pore-size filters. Scale bar: 150 nm.

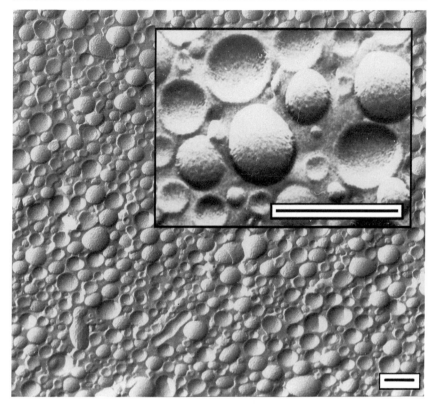

FIG. 5. Freeze–fracture micrograph of egg PC vesicles prepared at 400 mg/ml in 150 mM NaCl, 20 mM HEPES, pH 7.4, by extrusion through 100-nm pore-size filters. *Inset:* Magnified view of the vesicles. Scale bars: 200 nm.

proceeds rapidly and the rate is no longer temperature dependent. Surprisingly, in the presence of cholesterol (which abolishes the cooperative gel–liquid crystalline transition), the rate of extrusion below T_c is still slow (0.06 ml min^{-1} at 40°), whereas at 65° the extrusion rate is 200-fold higher. Similar effects were also observed for other saturated lipids. These results indicate that lipids in the gel state cannot be extruded at medium pressure and that extrusion rates at temperatures below the gel–liquid crystalline phase transition are prohibitively slow, even in the presence of cholesterol. It is reasonable to conclude that the inability to extrude below the phase transition temperature is most likely related to the much higher viscosity of gel-state membranes and their decreased deformability.[13] The observation that cholesterol slightly facilitates extrusion below T_c but reduces

[13] P. R. Cullis and B. de Kruijff, *Biochim. Biophys. Acta* **559,** 399 (1979).

extrusion rates above T_c correlates with the ability of cholesterol to decrease membrane viscosity below T_c and to increase viscosity above T_c. When saturated systems are extruded at temperatures at which the phospholipid is normally in a liquid crystalline state, size reduction and the formation of unilamellar vesicles proceed normally; however, users should be aware of some stability issues discussed below.

Liposomes composed of long-chain saturated lipids can be unstable when cooled below their T_c. For example, small vesicles produced by extrusion of DSPC or diarachidoyl phosphatidylcholine (DAPC) through filters with a pore size of 30 nm are metastable. The vesicles aggregate and fuse when incubated at 4 or 20°. This is likely due to the gel–liquid crystalline phase transition, which is associated with a large decrease in molecular surface area as lipid enters the gel state. This reduced surface area (which can be as much as 40 to 50%) is expected to destabilize vesicles. These effects can be observed by freeze–fracture when vesicles are prepared above the T_c and then cooled to below the T_c before cryofixing. Angular fracture planes are observed but not when cholesterol is present, consistent with its ability to prevent phospholipid from forming a cooperative, all-*trans* gel-state configuration, thus reducing changes in surface area as the temperature is decreased.[12]

For all practical purposes, extrusion of saturated systems is limited to lipids with a T_c below 100°. Successful extrusion of PCs with chain lengths ranging from 14 to 22 carbons has been achieved; the latter (dibehenoyl phosphatidylcholine) extrudes at 100° (M. J. Hope, unpublished data). Because of the high viscosity associated with membranes of long-chain saturated lipids, especially if extrusion occurs at or near the T_c, back pressure tends to be high and extrusion rates slow.

The majority of liposomes used either in drug delivery or as tools of membrane research are composed of phosphoglycerides or sphingomyelin and in our experience all liquid crystalline, bilayer-forming phospholipids (in isolation or as complex mixtures) are amenable to the extrusion technique. The rate of extrusion, or the operational pressure required to force liposomes through filters, varies with charge, acyl composition, pH, ionic strength, and the presence of interacting ions such as Ca^{2+} or Mg^{2+}. However, these factors do not usually prevent extrusion.

We (and others) have found that there is an advantage to extruding some liposome preparations in the presence of ethanol. Most lipids commonly used to prepare liposomes dissolve in this solvent and MLVs form spontaneously when the alcohol–lipid dispersion is diluted with buffer to a final ethanol concentration in the range of 10–25% (v/v). Not only is this ethanol–aqueous mixture readily extruded but the alcohol also facilitates the passage of lipid through the filter pores, resulting in lower back pressure and enhanced flow rate. The vesicles generated tend to exhibit an even more homogeneous size distribution around the pore size than is observed in the absence of alcohol. The ethanol concentration used generally does not affect the immediate permeability of the membrane to entrapped solute or can be conveniently diluted to such a level that it has no effect. Furthermore, ethanol is one of the few organic solvents acceptable in the manufacturing process of pharmaceutical products and its miscibility with water means that it is easily removed from vesicle preparations by gel filtration, dialysis, or tangential flow.

Applications

Finally, the extrusion process is readily scaled up to manufacture liposomes in large quantities for industrial and medical applications. The simplicity of the process means that complex equipment is not needed and sterility can be maintained. For example, research-scale equipment (Fig. 1) can be sterilized, depyrogenated, and operated in a sterile environment for drug delivery research. A scaled-up extrusion process can be accomplished in a number of ways, but the two most straightforward designs use either inert gas pressure, similar to the research-scale equipment described earlier, or a pump to drive liposome suspensions through in-line filter holders.

Extrusion is particularly effective at producing homogeneous vesicle populations with diameters from 70 to 150 nm, the most important range for liposome under development for intravenous administration. Vesicles of this size are small enough not only to circulate without becoming trapped in tissue microvasculature but also to accumulate at tumor and inflammation sites by extravasation through endothelial cell pores and gaps associated with these areas.[14,15] Furthermore, vesicles of this size have good drug-carrying capacity but are small enough to pass through sterilizing filters without damage.

[14] S. K. Hobbs, W. L. Mosky, F. Yuan, W. G. Roberts, L. Griffith, V. P. Torchilin, and R. K. Jain, *Proc. Natl. Acad. Sci. USA* **95,** 4607 (1998).
[15] S. K. Klimuk, S. C. Semple, P. Scherrer, and M. J. Hope, *Biochim. Biophys. Acta* **1417,** 191 (1999).

[2]   Giant Liposomes in Studies on Membrane Domain Formation

By Juha M. Holopainen, Miglena Angelova and Paavo K. J. Kinnunen

Introduction

There has been a renewal of interest in the organization of biomembranes. Accordingly, it has become well established that biomembranes are laterally highly heterogeneous, being organized into microdomains with specific lipid as well as protein compositions.[1] This ordering is due to mechanisms operating at thermal equilibrium as well as those involving an energy flux, that is, dissipative processes.[2] Several molecular mechanisms of the former category have been worked out and involve both lipid–protein as well as lipid–lipid interactions, with key contribution by the physicochemical properties of lipids.[3–5] Most of these studies have used liposomal model membranes together with various spectroscopic techniques. However, the diameters of liposomes obtained by methods such as extrusion are limited to ~100–200 nm, thus imposing a serious drawback when pursuing structures on the length scales relevant to plasma membranes, for example. This limitation was overcome by the introduction of so-called giant unilamellar vesicles, GUVs (see Luisi and Walde[6]), with diameters up to hundreds of microns. These model membranes have remained relatively little exploited. Yet, significant new information has been obtained about the physical properties of bilayers and shape transformations of vesicles, and in this context the pioneering studies by the groups of Evans,[7–9] Sackmann,[10,11] Needham,[12–14] and Bothorel[15] should be mentioned.

[1] P. K. J. Kinnunen, *Chem. Phys. Lipids* **57,** 375 (1991).

[2] P. K. J. Kinnunen, *Cell Physiol. Biochem.* **10,** 243 (2000).

[3] O. G. Mouritsen and P. K. J. Kinnunen, *in* "Biological Membranes" (K. Merz, Jr. and B. Roux, eds.), p. 463. Birkhäuser, Boston, 1996.

[4] J. Y. A. Lehtonen and P. K. J. Kinnunen, *Biophys. J.* **68,** 1888 (1995).

[5] J. M. Holopainen, J. Lemmich, F. Richter, O. G. Mouritsen, G. Rapp, and P. K. J. Kinnunen, *Biophys. J.* **78,** 2459 (2000).

[6] P. L. Luisi and P. Walde, eds., "Giant Vesicles." John Wiley & Sons, New York, 2000.

[7] R. Kwok and E. Evans, *Biophys. J.* **35,** 637 (1981).

[8] D. Needham, T. J. McIntosh, and E. Evans, *Biochemistry* **27,** 4668 (1988).

[9] D. Needham and E. Evans, *Biochemistry* **27,** 8261 (1988).

[10] J. Käs and E. Sackmann, *Biophys. J.* **60,** 825 (1991).

[11] H. G. Döbereiner, J. Käs, D. Noppl, I. Sprenger, and E. Sackmann, *Biophys. J.* **65,** 1396 (1993).

Importantly, GUVs can be directly observed by light microscopy. Moreover, they can also be subjected to micromanipulation techniques, including microinjection. Studies are now being published describing their use for the observation of membrane microdomains.[16,17] The inclusion of integral membrane proteins into GUVs has been described.[18] Clearly, giant liposomes hold great potential and will certainly be helpful in elucidating biomembrane properties and functions in a well-defined model system.

Several techniques have been described for the formation of GUVs. In early studies GUVs were made by first depositing the desired lipids in an organic solvent on a Teflon disk, the surface of which had been slightly roughened by sandpaper.[9] The disks were subsequently hydrated with water vapor and then immersed into an aqueous buffer. The yields are, however, rather modest and the procedure is time consuming. These problems are alleviated by the use of an AC electric field facilitating formation of the GUVs.[19] We describe this technique in detail in this chapter, together with its use to observe microdomain formation by fluorescence microscopy.

Giant Unilamellar Vesicle Electroformation Chamber

The electroformation of GUVs is best performed with a specialized chamber. One possible setup is illustrated in Fig. 1. The chamber consists of a circular cavity with a diameter of 26.5 mm, drilled in a metal plate, and with an opening with a diameter of 19 mm in its bottom. Into this chamber a cuvette (diameter, 26 mm) of optical quality quartz glass is fitted. Attached to the Plexiglas holder are two platinum electrodes (diameter, 0.8 mm), which can be removed for lipid deposition and cleaning. The distance between the parallel electrodes axes in the chamber is 4 mm.

---

[12] D. Needham and R. M. Hochmuth, *Biophys. J.* **55,** 1001 (1989).

[13] D. Needham and R. S. Nunn, *Biophys. J.* **58,** 997 (1990).

[14] D. V. Zhelev and D. Needham, *Biochim. Biophys. Acta* **1147,** 89 (1993).

[15] P. Meleard, C. Gerbeaud, T. Pott, L. Fernandez-Puente, I. Bivas, M. D. Mitov, J. Dufourcq, and P. Bothorel, *Biophys. J.* **72,** 2616 (1997).

[16] L. A. Bagatolli and E. Gratton, *Biophys. J.* **77,** 2090 (1999).

[17] J. Korlach, P. Schwille, W. W. Webb, and G. W. Feigenson, *Proc. Natl. Acad. Sci. USA* **96,** 8461 (1999).

[18] N. Kahya, E. I. Pecheur, W. P. de Boeij, D. A. Wiersma, and D. Hoekstra, *Biophys. J.* **81,** 1464 (2001).

[19] M. I. Angelova and D. S. Dimitrov, *Faraday Discuss. Chem. Soc.* **81,** 303 (1986).

Pt electrodes

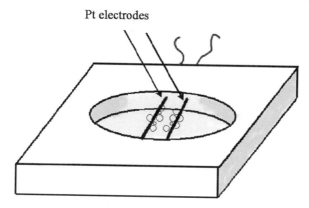

FIG. 1. A schematic diagram of one type of electroformation chamber.

Deposition of Lipids

The desired lipids are dissolved and mixed in an organic solvent at a concentration of 0.3 to 1.6 mg/ml. Of this solution, 1 to 3 μl is deposited with a microsyringe [Hamilton (Reno, NV) or an equivalent] in ~1-μl aliquots onto the platinum electrodes under a stream of nitrogen, in a manner allowing immediate evaporation of most of the solvent. Subsequently, the electrodes are maintained under reduced pressure for 2 to 24 h, so as to ensure complete removal of solvent residues. The electrodes are then fixed into the chamber inside the quartz cuvette. The formation of sphingomyelin/1 stearoyl-2-oleoyl-sn-glycero-3-phosphocholine (SOPC) (3:1, molar ratio) GUVs is described in detail in this chapter. For observation of the vesicles as well as to allow monitoring of the progression of enzymatic hydrolysis of sphingomyelin by sphingomyelinase, a fluorescent tracer, BODIPY-labeled sphingomyelin (mole fraction $X = 0.05$; Molecular Probes, Eugene, OR), is also included.

The chamber with the electrodes is then placed on the stage of an inverted fluorescence microscope, equipped with long or extralong working distance objectives and resting on a vibration isolation table (Melles Griot, Carlsbad, CA). Before hydration of the lipid an AC field (0.2–0.4 V, 4–10 Hz) is applied, using a voltage generator (CFG250; Tektronix, Beaverton, OH), and the chamber is then filled with buffer (1.3 ml of 0.5 mM HEPES, pH 7.4). This field is maintained for 1 min, after which the voltage is increased to 1.2–1.3 V and the frequency is adjusted to 4 Hz. The GUVs start to form on the electrodes and become visible by phase-contrast or fluorescence microscopy in about 0.5 h (Fig. 2). Initially, the formed GUVs are small and through subsequent fusions they grow in diameter. For easier

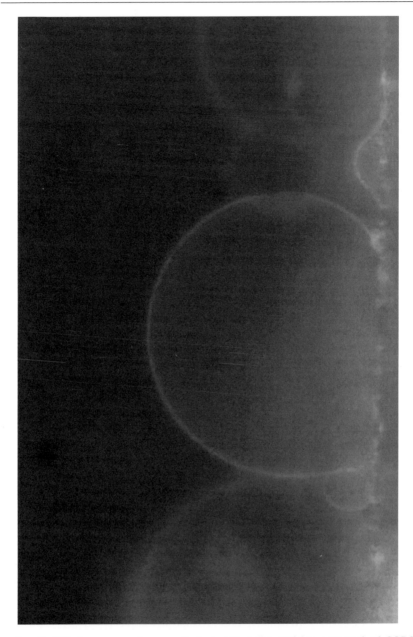

FIG. 2. Still fluorescence images of giant unilamellar vesicles composed of SOPC, N-palmitoyl–sphingomyelin, and BODIPY–sphingomyelin (0.75:0.20:0.05 molar ratio, respectively) after electroformation.

observation as well as micromanipulation GUVs attached to the electrode surface can be used. Alternatively, the GUV on the electrode may be allowed to become spherical before pulling it away from the electrode with a holding micropipette.[20] The number of bilayers in the GUV can be deduced when fluorescent tracer is present. Accordingly, the emission intensity of the liposome membrane is recorded with a charge-coupled device (CCD) camera and should increase as multiples. The bilayers with minimum values should represent unilamellar vesicles.[21] An alternative method is based on determining the elastic properties of the bilayer.[22]

Microinjection

One of the fascinating possibilities of GUVs is to subject them to local perturbation by specific lipid-modifying enzymes[23–26] or other membrane-active substances, such as DNAs[27] and antimicrobial peptides.[28] This is best achieved by microinjection. Micropipettes are pulled from a borosilicate capillary (outer diameter, 1.2 mm), using a programmable puller (Sutter P-87; Sutter Instruments, Novato, CA). Tip diameters are determined by measuring the threshold pressure required to obtain air bubble flow through the micropipette tip immersed into ethanol.[29] The micropipette is attached to a micromanipulator (MX831 with MC2000 controller; SD Instruments, Grants Pass, OR) and further connected by plastic tubing to a microinjector (PLI-100; Medical Systems, Greenvale, NY). Before loading the pipette with the enzyme solution the latter must be filtered in order to remove particles and aggregates, which could cause clogging of the micropipette tip. For this purpose the solution is passed through a 0.2-μm pore size filter (World Precision Instruments, Sarasota, FL). An aliquot (~100 μl) of the filtered enzyme solution is applied onto a clean microscope slide and the micropipette is immersed with the manipulator into the solution. The micropipette is filled by applying

[20] F. M. Menger and J. S. Keiper, *Curr. Opin. Chem. Biol.* **2,** 726 (1998).
[21] K. Akashi, H. Miyata, H. Itoh, and K. Kinosita, Jr., *Biophys. J.* **71,** 3242 (1996).
[22] M. I. Angelova, S. Soleau, P. Meleard, J. F. Faucon, and P. Bothorel, *Prog. Colloid Polym. Sci.* **89,** 127 (1992).
[23] R. Wick, M. I. Angelova, P. Walde, and P. L. Luisi, *Chem. Biol.* **3,** 105 (1996).
[24] V. Dorovska-Taran, R. Wick, and P. Walde, *Anal. Biochem.* **240,** 37 (1996).
[25] J. M. Holopainen, O. Penate Medina, A. J. Metso, and P. K. J. Kinnunen, *J. Biol. Chem.* **275,** 16484 (2000).
[26] J. M. Holopainen, M. I. Angelova, and P. K. J. Kinnunen, *Biophys. J.* **78,** 830 (2000).
[27] M. I. Angelova and I. Tsoneva, *Chem. Phys. Lipids* **101,** 123 (1999).
[28] H. Zhao, J. P. Mattila, J. M. Holopainen, and P. K. J. Kinnunen, *Biophys. J.* **81,** 2979 (2001).
[29] M. Schnorf, I. Potrykus, and G. Neuhaus, *Exp. Cell Res.* **210,** 260 (1994).

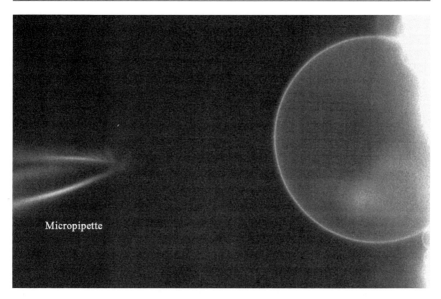

FIG. 3. A still fluorescence microscopy image of a micropipette close to a giant unilamellar vesicle composed of SOPC, N-palmitoyl–sphingomyelin, and BODIPY–sphingomyelin (0.75:0.20:0.05 molar ratio, respectively).

reduced pressure via the injector. Subsequently, the pipette is brought into the liposome electroformation chamber. Importantly, the view is easily calibrated by using proper multiples of the known step length of the micromanipulator or an object-micrometer. The pipette tip is then adjusted to the vicinity of the GUV surface (Fig. 3), for controlled delivery of the enzyme solution by the microinjector. In the shown study, the applied sphingomyelinase converts the sphingomyelin in the outer surface into ceramide, by hydrolytic cleavage of the phosphocholine head group. Unlike sphingomyelin, which is perfectly miscible with phosphatidylcholine, the produced ceramide readily segregates (within ~10 s) into brightly fluorescent microdomains (Fig. 4A). The latter is driven by intramolecular hydrogen bonding.[30] Subsequently, the domains invaginate and form "endocytotic-like" vesicles into the inner space of the GUV (Fig. 4B).

One of the interesting possibilities available when using GUVs is microinjection into vesicle inner space. This requires a thin, long micropipette tip (inner/outer diameter, 0.1/0.2 μm), and the use of a vibration isolation table. As expected, a number of vesicles burst in the process of inserting

[30] I. Pascher, *Biochim. Biophys. Acta* **455**, 433 (1976).

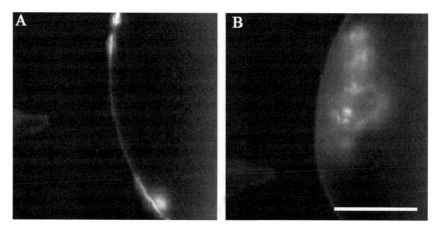

FIG. 4. Sphingomyelinase (*Bacillus cereus*) was applied in the vicinity of the outer membrane of a GUV composed of SOPC, *N*-palmitoyl–sphingomyelin, and BODIPY–sphingomyelin (0.75 : 0.20 : 0.05 molar ratio, respectively). In a few tenths of seconds brightly fluorescent spots appeared in the membrane (A). On further incubation these spots were invaginated as smaller vesicles into the interior of the GUV (B). Note that only a small portion of the GUV is shown. Scale bar in (B) corresponds to 100 μm.

the micropipette through the bilayer. However, the success rate is reasonable. A GUV with a micropipette tip inside of it is shown in Fig. 5.

Aspects to Be Aware of

It should be emphasized that although some theoretical possibilities have been proposed,[31] the exact mechanisms of liposome electroformation are still not understood entirely. Each electroformation case, that is, any particular lipid composition and buffer, needs to be considered individually and takes a few trials before setting an efficient protocol. This is not a complicated process, because the operating person can observe the vesicle formation directly and control the processes.

The vesicles obtained by electroformation are spherical and when attached to the electrodes lack thermal undulations. Accordingly, the bilayer is under small yet finite tension. The impact of the latter on the bilayer properties is not known yet. If well-isolated and relaxed GUVs are demanded, the suspension should be taken (gently) out of the preparation chamber and transferred into another working chamber. However, once the conditions have been worked out the electroformation method is highly reproducible and fast. The choice of lipids warrants consideration. As

[31] M. I. Angelova and D. S. Dimitrov, *Prog. Colloid Polymer Sci.* **76**, 59 (1988).

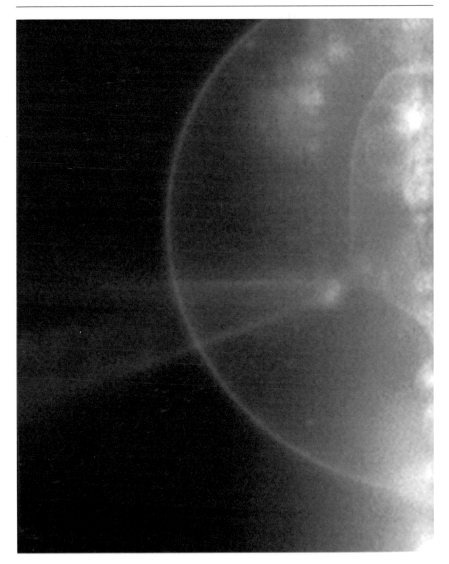

Fig. 5. A still fluorescence image of a GUV composed of SOPC, N-palmitoyl–sphingomyelin, and BODIPY–sphingomyelin (0.75:0.20:0.05 molar ratio, respectively) showing a micropipette embedded in the interior of the vesicle.

expected on the basis of augmented van der Waals interactions,[4] more stable GUVs are formed when using lipids with longer saturated acyl chains. For instance, SOPC is preferred over 1-palmitoyl-2-oleoyl-*sn*-glycero-3-phosphocholine (POPC). When fluorescence microscopy or spectroscopy studies on GUVs are to be performed, the choice of fluorescent moiety is important. Obviously, the quantum yield of the fluorophore should be high and it should be stable against photobleaching. In addition, a small fluorescent moiety is preferable in order to reduce steric perturbation of the lipid packing in the GUV membrane.

The most critical factors in the efficient formation of GUVs are the thickness of the initial dried lipid film, and the applied voltage and frequency. Likewise, the duration of the exposure to the AC field varies and must be found experimentally. Some useful values can be found in Fischer et al.[32] and can be used as guidelines when initiating studies with a new system.

[32] A. Fischer, P. L. Luisi, T. Oberholzer, and P. Walde, in "Giant Vesicles" (P. L. Luisi and P. Walde, eds.), p. 37. John Wiley & Sons, New York, 1999.

[3] Preparation and Quantitation of Small Unilamellar Liposomes and Large Unilamellar Reverse-Phase Evaporation Liposomes

By NEJAT DÜZGÜNEŞ

Introduction

Since the publication of the first article on the preparation and characterization of multilamellar liposomes,[1] numerous methods have been developed to generate liposomes of different size and characteristics. Here we describe the preparation of two types of liposomes: (1) small unilamellar liposomes prepared by bath sonication, (2) large unilamellar liposomes prepared by reverse-phase evaporation.

Preparation of Small Unilamellar Liposomes

Small unilamellar liposomes were developed by Papahadjopoulos and Miller[2] and characterized thoroughly by Huang[3] and Suurkuusk et al.[4] For a typical preparation, for example, to encapsulate a fluorescent

[1] A. D. Bangham, M. M. Standish, and J. C. Watkins, J. Mol. Biol. 13, 238 (1965).
[2] D. Papahadjopoulos and N. Miller, Biochim. Biophys. Acta 135, 624 (1967).
[3] C. Huang, Biochemistry 8, 344 (1969).
[4] J. Suurkuusk, B. R. Lentz, Y. Barenholz, R. L. Biltonen, and T. E. Thompson, Biochemistry 15, 1393 (1976).

marker for membrane permeability studies or for monitoring liposome–cell interactions, 10 μmol of a phospholipid mixture in chloroform is placed in a glass tube (e.g., a high-quality Kimax glass tube (Fischer Scientific, Santa Clara, CA or Springfield, NJ)), ~1 cm in diameter, with a "screw cap" with Teflon lining at the top, using glass Hamilton syringes. Any fluorescent lipids to be incorporated in the liposome membrane are included in the chloroform mixture at this stage. Teflon tape is "sealed" over the top of the tube, the tube is placed in a larger glass tube with a fitting appropriate for a rotary evaporator (Büchi, with attached vacuum gauge and purge tubing connected to an argon tank), and the lipid is evaporated to dryness by constant, medium-speed rotation. The length of the process depends on the level of vacuum achieved. The evaporator is then purged with argon to reduce the vacuum, and the tubes are removed from the evaporator. The smaller tube is then placed in a vacuum oven at room temperature for at least 2 h to evaporate further any remaining chloroform.

The dried film is hydrated with an appropriate solution to be encapsulated and the tube is purged with argon gas (10–15 s), preferably with a syringe filter (0.22 μm) at the end of the tubing carrying the gas, sealed with Teflon tape, and capped immediately. The tube is mixed vigorously with a vortex mixer for 10 min at room temperature, or at a temperature above the gel-to-liquid crystalline phase transition of the lipid mixture. The latter may be achieved by intermittently placing the tube in a water bath about 10° above the phase transition temperature. The resulting multilamellar suspension is sonicated in a bath-type sonicator (Laboratory Supply, Hicksville, NY) for 0.5–1 h. It is important to ascertain that the water in the middle of the bath is breaking up into droplets due to the sonication, and that the top of the liposome suspension is at the same level as the water in the bath. The mixture in the tube must be observed to be agitating vigorously, with an aerosol forming occasionally in the argon layer above. If the sonicator is not producing sufficient energy, it may be necessary to replace the transducer at the bottom of the bath, or to adjust the power supply after consulting with the manufacturer. Overheating of the bath must be avoided. It is preferable to maintain the bath at room temperature either by circulating water through the bath (while keeping the water level steady), or by adding ice intermittently and readjusting the water level. The bath temperature must be maintained above the phase transition temperature of the lipids in the mixture.

The resulting suspension may be opalescent or clear, depending on the lipids being used. The suspension is centrifuged at 100,000g for 1 h, preferably in a swinging bucket rotor, to eliminate any remaining large liposomes. The supernatant is used as the small unilamellar liposome preparation.

Preparation of Large Unilamellar Liposomes by Reverse-Phase Evaporation

The reverse-phase evaporation technique was developed by Szoka and Papahadjopoulos,[5] and refined further.[6,7] The method enables the preparation of large unilamellar liposomes with a large capture volume. The primary drawback of the method is the use of diethyl or isopropyl ether, which precludes large-scale preparation.

The mixture of phospholipids to be used, including any fluorescent lipids, is dissolved in chloroform (total lipid, 10–20 μmol) in a high-quality glass tube (which will not break during the sonication step), as described previously. The phospholipids to be used should be measured and transferred with glass Hamilton syringes. Teflon tape is sealed over the top of the tube, and the tube is placed in a larger glass tube with a fitting appropriate for a rotary evaporator and dried in vacuum.

A few milliliters of diethyl (or isopropyl) ether are washed with a similar volume of distilled or purified water in a tightly capped glass tube with a Teflon fitting, by gentle shaking under a chemical hood. Isopropyl ether is used for lipids with a high gel–liquid crystalline transition temperature (T_m), because its boiling point is higher than that of diethyl ether, and the eventual evaporation step takes place above the T_m. The mixture is allowed to settle. One milliliter of the ether (the top layer) is added to the dried phospholipid film, using a glass pipette or syringe. The tube is agitated gently to ensure that the lipid film dissolves in the ether.

The buffer to be encapsulated (0.34 ml) is added to the phospholipid solution in ether. The tube is flushed with a stream of argon gas connected via Tygon tubing to a Pasteur pipette immobilized with a clamp on a stand, and the tube is sealed with Teflon tape and the screw cap. The mixture is sonicated in a bath-type sonicator (Laboratory Supply) for 2–5 min, ensuring that the surface of the mixture breaks up into small droplets. The ether and aqueous phases should not separate and should form a stable emulsion. After opening the screw cap, preferably under a hood, the tube is sealed again with Teflon tape and placed inside the larger glass tube that fits onto

[5] F. C. Szoka, Jr. and D. Papahadjopoulos, *Proc. Natl. Acad. Sci. USA* **75,** 4194 (1978).

[6] F. Szoka, F. Olson, T. Heath, W. Vail, E. Mayhew, and D. Papahadjopoulos, *Biochim. Biophys. Acta* **601,** 559 (1980).

[7] N. Düzgüneş, J. Wilschut, R. Fraley, and D. Papahadjopoulos, *Biochim. Biophys. Acta* **642,** 182 (1981).

the rotary evaporator. About 1 ml of water is included in the outer tube, both to maintain thermal contact and to minimize evaporation of the aqueous solution in the inner tube. The outer tube is immersed in the water bath of the rotary evaporator, maintained at 30°.

The ether is evaporated in controlled vacuum (~350 mmHg) under constant supervision (because the bubbles formed in the emulsion can start creeping up the tube if the vacuum is increased further). The suspension is purged occasionally with argon attached to the evaporator to maintain the vacuum level and to prevent excessive bubbling. When most of the ether has evaporated a gel is formed. At this point the vacuum is allowed to increase. After a few minutes the inner glass tube is removed and mixed vigorously on a vortex mixer for 5–10 s to break up the gel. The tube is placed again in the outer tube and rotary evaporation is resumed. This step is repeated once or twice, until an aqueous opalescent suspension is formed. At this stage, an additional 0.66 ml of the encapsulation buffer is added to the suspension and rotary evaporation is continued for an additional 20 min to remove any residual ether.

The liposome suspension is passed numerous times through polycarbonate membranes of 0.1 μm or other pore diameter (Poretics, Pleasanton, CA), using a high-pressure extruder (Lipex Biomembranes, Vancouver, BC, Canada) or syringe extruder (Avestin, Ottawa, ON, Canada; Avanti Polar Lipids, Alabaster, AL) to achieve a uniform size distribution.[6,8] Two stacked membranes separated by a drain disk may be used with the high-pressure extrusion device, and the liposome suspension may be passed through the membranes five times. The syringe extruders are more convenient for multiple extrusions, and the liposomes may be extruded 21 times.[9]

Phosphate Assay to Determine Phospholipid Concentration of Liposomes

After preparation and possible column chromatography to eliminate unencapsulated material, liposomes may be quantified by measuring their phosphate content, as long as the liposomes contain at least some phospholipid. We have utilized the original method described by Bartlett,[10] with

[8] N. Düzgüneş, J. Wilschut, K. Hong, R. Fraley, C. Perry, D. S. Friend, T. L. James, and D. Papahadjopoulos, *Biochim. Biophys. Acta* **732**, 289 (1983).
[9] S. Simões, V. Slepushkin, R. Gaspar, M. C. Pedroso de Lima, and N. Düzgüneş, *Gene Ther.* **5**, 955 (1998).
[10] G. R. Bartlett, *J. Biol. Chem.* **234**, 466 (1959).

modifications introduced by T. D. Heath and D. Alford in the laboratory of the late D. Papahadjopoulos (see Düzgüneş et al.[11]).

The samples to be tested are placed at the bottom of disposable borosilicate glass tubes (18 × 150 mm) in triplicate at an estimated amount of less than 0.1 μmol of inorganic phosphate. Phophate standards (e.g., 0.01, 0.05, 0.075, and 0.1 μmol of inorganic phosphate; Sigma, St. Louis, MO) are also pipetted in triplicate. Sulfuric acid (0.4 ml of a 10 N solution) is added to each tube from a polypropylene dispenser jar, and the tubes are placed for 30 min in a 160–170° heating block with inserts for the test tubes.

The tubes are then taken out of the heating block and allowed to cool. H_2O_2 (0.1 ml of a 9% solution) is added to each tube, using a pipettor or repeater pipette (Pipetman or Eppendorf). The tubes are heated again for 30 min on the heating block. The fumes arising from the tubes are tested for the absence of H_2O_2, using indicator strips (EM Quant peroxide test; EM Science, Gibbstown, NJ). The tubes are then cooled, and 4.6 ml of a 0.22% ammonium molybdate reagent in 0.25 N H_2SO_4 is dispensed into each tube, using an Oxford jar pipettor. The tubes are mixed thoroughly with a vortex mixer. To each tube 0.2 ml of ANSA (or Fiske) reagent is added, using a repeater pipette, and the tubes are vortexed. ANSA reagent is prepared as follows: 250 mg of aminonaphtholsulfonic acid and 500 mg of Na_2SO_3 are added to a 15% $NaHSO_3$ solution, bringing the volume up to 100 ml; the mixture is heated gently on a stir plate to dissolve all ingredients. Alternatively, 0.1 ml of ascorbate (15 g of ascorbic acid is dissolved in 100 ml of distilled or purified water and the solution is stored in the cold) can be used for this step. The tubes are placed in a metal rack that can fit into a boiling water bath, and incubated for 7–10 min. The rack is then placed in a cold water bath to cool.

The contents of the tubes are transferred carefully (e.g., wearing latex gloves and plastic goggles) to disposable spectrophotometer cuvettes, or a spectrophotometer with a sipper accessory can be utilized. The samples are read at 812 nm, or at 660 nm if the solution is highly concentrated. A calibration curve (usually a straight line) is generated from the readings of the phosphate standards, and the phosphate content of the liposome sample is determined on the basis of this curve.

[11] N. Düzgüneş, L. A. Bagatolli, P. Meers, Y.-K. Oh, and R. M. Straubinger, in "Liposomes: A Practical Approach" (V. Weissig and V. Torchilin, eds.), 2nd Ed., pp. 105–147. Oxford University Press, Oxford, 2003.

[4] Liposomes Prepared by High-Pressure Homogenizers

By RAMON BARNADAS I RODRÍGUEZ and MANUEL SABÉS I XAMANÍ

Introduction

High-pressure homogenizers are used for the preparation of liposomes and lipid dispersions because of their vesicle disruption capability.[1–7] In general, the initial suspension to be processed is composed of multilamellar liposomes that will be downsized inside the device by applying a large amount of energy to the suspension. The sample is injected at high and constant pressure in a specially designed part of the homogenizer where the rearrangement of liposome structure takes place due to turbulence, cavitation, and/or shear phenomena. The characteristics of the section where homogenization occurs represent one of the main differences between the models provided by the principal manufacturers.[8] Some have a dynamic valve, whereas others have a fixed geometry, and still others are equipped with dynamic or static homogenizing valves. In all cases, the maximum process pressures reached by these instruments (normally about 30,000 lb/in^2, but, in special cases, up to 60,000 lb/in^2) allow them to homogenize samples with a phospholipid concentration higher than 150 mg/ml. Another advantage associated with these devices is that, in most cases, small models have their corresponding scaled-up homogenizers for large-scale production and, subsequently, results can be transferred directly from laboratories to industry.

Properties of liposomes prepared by high-pressure homogenization depend on a first set of parameters related to the homogenizer and a

[1] E. Mayhew, R. Lazo, W. J. Vail, J. King, and A. M. Green, *Biochim. Biophys. Acta* **775**, 169 (1984).

[2] C. Washington, *Manufacturing Chemist* **49** (March 1988).

[3] H. Talsma, A. Y. Özer, L. van Bloois, and D. J. A. Crommelin, *Drug. Dev. Ind. Pharm.* **15**, 197 (1989).

[4] S. Vemuri, C. D. Yu, V. Wangsatorntanakum, and N. Roosdorp, *Drug Dev. Ind. Pharm.* **16**, 2243 (1990).

[5] M. Brandl, D. Bachmann, M. Drechsler, and K. H. Bauer, *Drug Dev. Ind. Pharm.* **16**, 2167 (1990).

[6] D. Bachmann, M. Brandl, and G. Gregoriadis, *Int. J. Pharm.* **91**, 69 (1993).

[7] S. Liedtke, S. Wissing, R. H. Müller, and K. Mäder, *Int. J. Pharm.* **196**, 183 (2000).

[8] http://www.apv.com/; http://www.avestin.com/; http://www.microfluidicscorp.com/

second group of factors associated with the sample. In the case of a given device with a specific homogenizing piece, the pressure and number of times that the sample is processed clearly determine the size distribution of the liposomes obtained. Sample factors include aspects such as phospholipid composition and concentration, initial size distribution and lamellarity of the liposomes, temperature, and composition and ionic strength of the bulk medium. In this chapter we describe procedures for obtaining liposomes and proteoliposomes (with a membrane protein), using a high-pressure homogenizer with a homogenizing piece having no moving parts, and the effect of some of the factors that influence vesicle size distribution.

Materials and General Procedures

Phospholipid Sources

All soybean phospholipids are purchased from Lucas Meyer (Hamburg, Germany). Emulmetik 930 is a deoiled, phosphatidylcholine-enriched fraction of soybean lecithin. It contains a minimum of 97% phospholipids, mainly phosphatidylcholine [minimum, 72% (w/w)], phosphatidylethanolamine [minimum, 8% (w/w)], phosphatidylinositol (maximum, 1%), and lysophosphatidylcholine [maximum, 3% (w/w)]. Emulmetik 950 is a hydrogenated soybean lecithin. It contains hydrogenated phosphatidylcholine [minimum, 95% (w/w)], lysophosphatidylcholine [maximum, 1% (w/w)], other phospholipids [maximum, 2.5% (w/w)], and oil [maximum, 1.0% (w/w)]. Pro-Lipo-S is a mixture of phosphatidylcholine and other soybean phospholipids (30%, w/w) as well as a hydrophilic medium (water, ethanol, and glycerol). This mixture mainly forms stacked, negatively charged bilayers that, when mixed with aqueous medium by stirring at room temperature, convert into liposomes.[9] Egg yolk phosphatidylcholine is purified according to the method described by Singleton et al.[10]

Purification of Bacteriorhodopsin

Purple membrane containing bacteriorhodopsin is obtained from Halobacterium salinarum as described by Oesterhelt and Stoeckenius.[11] The membrane sheets isolated have a lipid-to-protein ratio of 1:3 (w/w).

---

[9] S. Leigh, European patent application, application number 85301602.0; publication number, 0 158 441 (1985).

[10] W. S. Singleton, M. S. Gray, M. L. Brown, and J. L. White, J. Am. Oil Chem. Soc. 42, 53 (1965).

[11] D. Oesterhelt and W. Stoeckenius, Methods Enzymol. 31, 667 (1974).

High-Pressure Homogenization

A Microfluidizer 110S (Microfluidics, Newton, MA) is utilized to prepare liposomes by high-pressure homogenization. In this laboratory-scale model, homogenization pressure is 230 times the inlet pressure. This device is equipped with a ceramic interaction chamber with fixed geometry where homogenization takes place. When the sample is processed inside, the flow splits into two main streams. They are forced to impact with one another at great velocity before leaving the interaction chamber. Depending on the characteristics of the sample and of the pressure, this recombination results in a specific reduction of the size of the vesicles present in the suspension. When working with slurried and/or concentrated suspensions at low pressures, the interaction chamber can become plugged. In this case, the chamber can be cleared easily by reversing its position in order to back flush. As a result of the position of a spool valve, the Microfluidizer 110S can operate by recirculating the processed sample to a product inlet reservoir or in a nonrecirculating mode. A removable coil and bath allow, when necessary, control of the temperature of the sample immediately before processing inside the interaction chamber.

Determination of Liposome Size

The size distribution of liposomes is measured by dynamic light scattering, using an ultrafine particle analyzer (UPA) 150 spectrometer (Microtrac, Montgomeryville, PA). This device operates by means of heterodyne detection[12,13] and, for mathematical modeling, it assumes that only Brownian motion produces the velocity distribution of the particles. The spectrometer is equipped with a diode laser having a wavelength of 780 nm, and has an optical power of 3 mW. Analysis acquisition time is 10 min, and the samples are diluted with their aqueous medium to obtain a satisfactory signal in the detector. Results are presented as volume (or mass) distribution and are expressed as the mean diameter and width (half the central range of the measured particle size distribution that contains 68% of the vesicles).

Liposome Homogenization in Nonrecirculation Mode

Principle

Because of the constant pressure applied to the sample, large-scale preparation of liposomes with high-pressure homogenizers is highly reproducible. Therefore, at a given pressure, liposome size distribution

---

[12] N. Ostrowsky, *Chem. Phys. Lipids* **64,** 45 (1993).
[13] M. N. Trainer, P. J. Freud, and E. M. Leonardo, *Am. Lab.* **37,** 34 (1992).

depends on the number of times that vesicles pass through the interaction chamber, and on the characteristics of their own suspension.[14] When a non-recirculating mode of operation is selected, all the suspension undergo the same process, and, consequently, the times (number of cycles) that all the liposomes are processed clearly determine the final diameter of the vesicles. On the other hand, as the bilayer charge influences liposome size during their formation,[15,16] the ionic strength of the bulk medium is an important parameter to control, in order to regulate the bilayer potential and, as a result, vesicle size distribution.

Methods

To obtain the initial liposome raw suspension, sodium phosphate buffer (10 mM, pH 7.4) is poured into and mixed with Pro-Lipo-S (the Steward assay can be employed[17] in order to determine the final phospholipid concentration, thus avoiding buffer interference), or, alternatively, it is possible to use a non-phosphate-containing buffer (e.g., HEPES). The ionic strength (IS) of the aqueous medium is adjusted with required quantities of NaCl and the phospholipid concentration is kept constant in all samples and equal to 50 mg/ml. All homogenizations are carried out at room temperature. To study the effect of cycles (C, ranging from 1 to 9), inlet pressure (p, ranging from 0.8 to 4 atm), and ionic strength (ranging from 22 to 155 mM) on liposome size distribution, a central composite experimental design[18] is used. The combination and independent replicate factors of this design are shown in Table I. After processing the samples under the desired conditions, the size distribution of vesicles is obtained with the UPA 150. The relationship between the factors (C, p, and IS) and the responses (mean diameter and width) is calculated by the stepwise method, fitting empirical, full second-order polynomial models that include constant, first-order, second-order, and interaction terms. In these equations, the factor levels are expressed in coded values ranging from -2 to 2. This procedure allows the estimated values of the empirical parameters not to depend on each other and to facilitate the matrix manipulations.[18] Consequently, the general expression of the equations is

[14] R. Barnadas and M. Sabés, *Int. J. Pharm.* **213,** 175 (2001).

[15] D. D. Lasic, "Liposomes: From Physics to Applications," Chapter 3. Elsevier Science, Amsterdam, 1993.

[16] K. Akashi, H. Miyata, H. Itoh, and K. Kinosota, *Biophys. J.* **74,** 2973 (1998).

[17] R. R. C. New, "Liposomes: A Practical Approach." IRL Press, Oxford, 1990.

[18] S. N. Deming and S. L. Morgan, "Experimental Design: A Chemometric Approach." Elsevier Science, Amsterdam, 1987.

$$\text{Response} = \text{constant} + \alpha_1 C^* + \alpha_2 p^* + \alpha_3 \text{IS}^* + \beta_1 C^{*2} + \beta_2 p^{*2}$$
$$+ \beta_3 \text{IS}^{*2} + \gamma_1 C^* p^* + \gamma_2 C^* \text{IS}^* + \gamma_3 p^* \text{IS}^* \tag{1}$$

where α_1, β_1, and γ_1 are the empirical parameters calculated by the stepwise method, and

$$C^* = \frac{C - 5}{2} \tag{2}$$

$$p^* = \frac{p - 2.4}{0.8} \tag{3}$$

$$\text{IS}^* = \frac{\text{IS} - 88}{33} \tag{4}$$

After correlations are realized, the final equations contain only the significant parameters.

TABLE I
EXPERIMENTAL DESIGN USED TO STUDY NON-RECIRCULATING HOMOGENIZATION [a]

Cycles		Inlet pressure		Ionic strength		
Absolute	Coded	Absolute (atm)	Coded	Absolute (mM)	Coded	n
1	−2	2.4	0	88	0	3
5	0	0.8	−2	88	0	3
5	0	2.4	0	22	−2	3
5	0	2.4	0	88	0	3
5	0	2.4	0	155	2	3
5	0	4	2	88	0	3
9	2	2.4	0	88	0	3
3	−1	1.6	−1	55	−1	1
3	−1	1.6	−1	121	1	1
3	−1	3.2	1	55	−1	1
3	−1	3.2	1	121	1	1
7	1	1.6	−1	55	−1	1
7	1	1.6	−1	121	1	1
7	1	3.2	1	55	−1	1
7	1	3.2	1	121	1	1

[a] Factor combinations and replicates of the experimental design used to study the effect of pressure, cycles, and ionic strength on liposome size distribution obtained with the Microfluidizer 110S. Factor levels are indicated in absolute and coded values. Reprinted from *International Journal of Pharmaceutics*, **213**, R. Barnadas and Manuel Sabés, "Factors involved in the production of liposomes with a high-pressure homogenizer," 175–186 (2001), with permission from Elsevier Inc.

Results

The fitted empirical equations obtained [Eqs. (5) and (6)] are shown in Table II. Both models pass the statistical test for the effectiveness of the factors, have good coefficients of multiple correlation, and have low standard errors of estimate. The slopes of the surface responses are significantly dependent on the three factors. Some examples of surface responses are shown in Fig. 1. In the studied range, pressure has a continuous effect on liposome size (Fig. 1A and B) as any increase in pressure causes a decrease in liposome diameter, although at high pressures the slope of mean diameter and width curves tend to zero. In the case of cycles, however, any increment in the number of cycles larger than approximately 7 does not significantly decrease the mean diameter or the width. Although not as strong as pressure and cycles, ionic strength also has an appreciable effect on liposome size distribution (Fig. 1C and D). As a result of the screening of bilayer electrical charges and, therefore, a diminution of the interbilayer repulsion, any increase in ionic strength causes an increase in liposome size.

From the point of view of the modality of the liposome suspensions, homogenization of Pro-Lipo-S yields mainly bimodal populations of vesicles. But from the surface responses and from the size distribution results, it is possible to predetermine the necessary conditions for obtaining two different unimodal samples. First, by processing samples for 9 cycles at 4 atm of inlet pressure and an ionic strength of 22 mM, small vesicles are obtained with a mean diameter of 39 ± 7 and 15 ± 4 nm in width ($n = 3$). From Eq. (5) and Eq. (6) in Table II, the estimated values obtained with the previous factor levels are, correspondingly, 62 and 73 nm. Conditions to obtain the second unimodal suspension are attained by taking into account the evolution of the ratio between the liposome

TABLE II

EFFECT OF PRESSURE, CYCLES, AND IONIC STRENGTH ON LIPOSOME SIZE DISTRIBUTION[a]

Equation no.	Response	R^2	F_{calc} ($F_{crit} = 2.64$)	Standard error of estimate
Eq. (5)	Mean diameter (nm) = $151 - 49.3\,C^* - 55.7p^* + 15.8\,IS^* + 21.9\,C^{*2} + 16.3p^{*2}$	0.982	125	18.6
Eq. (6)	Width (nm) = $129 - 57.7\,C^* - 82.6p^* + 9.34\,IS^* + 30.7\,C^{*2} + 30.1p^{*2}$	0.983	130	25.0

[a] Fitted empirical equations of mean diameter and width and their statistical results. The studied factors are the inlet pressure of the Microfluidizer, the number of cycles, and the ionic strength of the bulk medium.

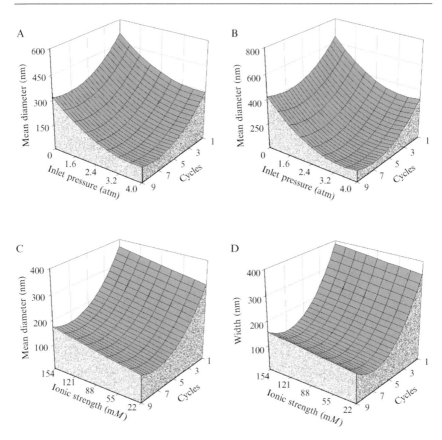

FIG. 1. Surface responses of the mean diameter and width of the size distribution of liposomes obtained with the Microfluidizer 110S. In the case of (A) and (B), the ionic strength is 88 mM. In the case of (C) and (D), the inlet pressure is 2.4 atm.

populations during homogenization. For this purpose, factor levels needed are 1 cycle, 2 atm of inlet pressure, and an ionic strength of 22 mM. With these settings, the experimental mean diameter is 319 ± 6 nm, with a width of 83.2 ± 13.4 nm ($n = 3$). In this case, the values predicted by the model are 338 nm in the case of the mean diameter and 397 nm in the case of the width. Observe that in both unimodal suspensions, the measured mean diameters are comparable with the estimated values, when considering the standard error of estimate (Table II) and the experimental variability. This does not occur in the case of width. This fact can be explained by taking into account that the surface responses are obtained mainly from bimodal samples. During homogenization, the experimental mean diameter (a measure of central tendency) will always decrease as a consequence of vesicle size diminution and, consequently, the model will predict the mean diameter of both the unimodal and bimodal samples. On the other hand, the width (which depends on the dispersion of the samples) can show diminutions and increments during the vesicle downsizing if unimodal and bimodal samples are obtained during the process. In the case of samples processed for 1 cycle at 2 atm and at an ionic strength of 22 mM, the corresponding region of the width surface has been obtained from bimodal samples, and, consequently, the predicted value is considerably different from the experimental value (a local minimum is not included in the equation). In the case of samples processed for 9 cycles at 4 atm and at an ionic strength of 22 mM, the difference between the experimental and the estimated width is not as high as in the previous case. This could be caused by the fact that, in this region, the samples used to calculate the width surface are already mainly unimodal.

Liposome Homogenization in Recirculation Mode

Principle

When the Microfluidizer operates in the recirculating mode, the processed parts of the suspension are mixed with the sample contained in the reservoir. Therefore, the system is similar to a continuous stirrer tank reactor with the same inlet and outlet flow rates, where mixing takes place between liposomes that have passed a different number of times through the interaction chamber. With a sufficient period of time, all liposomes experience a disruption and, if the sample is extensively homogenized, vesicles reach the minimum diameter value allowed by the pressure of processing and by sample characteristics. For a constant flow rate (in the case of homogenizers depending on the pressure), the time evolution of this type of system depends on the volume of the sample being processed.

Method

Liposome suspensions are obtained by pouring and mixing sodium phosphate buffer (10 mM, pH 7.4) with Pro-Lipo-S. The phospholipid concentration of all samples is 50 mg/ml and they are processed at 4 atm of inlet pressure at room temperature. The spool valve is selected in recirculating mode and the sample volumes processed are 15, 30, 45, 60, and 90 ml (in all cases $n = 3$). As the maximum volume of the sample reservoir is 25 ml, a 400-ml sample reservoir is installed when needed. In these cases, in order to produce optimal mixing in the reservoir, mechanical stirring

TABLE III
LIPOSOMES OBTAINED IN RECIRCULATION MODE[a]

Sample volume (ml)	$A \pm SD$ (nm)	$\tau \pm SD$ (s)	R^2
15	1087 ± 24	16.81 ± 0.26	0.9714
30	1062 ± 37	33.2 ± 3.91	0.9075
45	1099 ± 24	59.5 ± 3.80	0.9626
60	1079 ± 18	81.9 ± 4.35	0.9832
90	1051 ± 26	104 ± 8.6	0.9282

[a] Fitted parameters of the exponential decay of the mean diameter when liposomes are processed with recirculation in the Microfluidizer 110S (value ± standard deviation; $n = 3$).

is applied internally by means of flat plates. At specific time intervals, aliquots of 0.2 ml are taken from the sample reservoir and are analyzed with the UPA 150. Maximum homogenizing times range from 4 min, in the case of 15-ml samples, to 10 min in the case of 90-ml samples.

Results

During homogenization, the mean diameter diminishes in all samples over time until reaching a constant value. This value, 28 ± 7 nm ($n = 15$), is significantly different ($p < 0.05$) from that obtained through processing the same type of sample for 9 cycles at 4 atm (39 ± 7 nm; see previous results) but shows neither a practical nor statistically significant difference if their corresponding widths of size distribution are taken into account (13 ± 4 and 15 ± 4 nm, respectively). Therefore, the constant mean diameter reached in all cases corresponds to the minimum vesicle diameter that can be obtained at the operating pressure. All results have a good fit to a time exponential decay (Table III), with the next general expression being

$$\text{Mean diameter (nm)} = d_m + A \exp(-t/\tau) \qquad (7)$$

where t is time in seconds, d_m equals 28 nm (this is not allowed to vary during the fitting procedure and, consequently, becomes the horizontal asymptote), and A (nm) and τ (s) are the parameters specified by the fitting procedure.

Parameter τ is equivalent to the residence time used in describing time evolution in continuous stirrer tank reactors and, as shown in Fig. 2, has a good correlation with the sample volume. Because of this fact, and considering that the curve passes through the coordinate origin, all equations

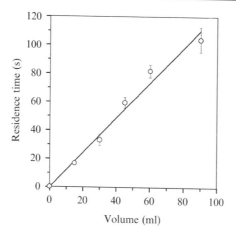

FIG. 2. Relation between residence time (τ) and the volume of the sample processed in the Microfluidizer 110S in recirculation mode. The value of τ allows fitting the measured mean diameter of the liposomes to the time of sample recirculation.

describing mean diameter variation [Eq. (7)] can be expressed as a function of time/volume, becoming

$$\text{Mean diameter (nm)} = d_m + A \, \exp\left(-\frac{t}{1.242v}\right) \tag{8}$$

where V is the sample volume in milliliters.

Experimental results using Eq. (8) are shown in Fig. 3A. The absolute time scale is obtained by multiplying the x axis values by the volume of the processed sample.

The width distribution shows similar behavior to the mean diameter (Fig. 3B). For any time greater than a critical value, the width diminishes until it becomes constant (13 ± 4 nm; $n = 15$). This width is not significantly different ($p > 0.05$) from the minimum width obtained through processing the sample for 9 cycles at 4 atm and at an ionic strength of 22 mM (15 ± 4 nm; see previous results). As in the case of the mean diameter, all experiments show a comparable evolution when represented as a function of t/V (Fig. 3B). At the outset of homogenization, all samples show an increment of the width because of the mixing of small quantities of processed sample with the suspension contained in the sample reservoir. Consequently, the width increment reflects an initial increment of the vesicle size distribution. After a maximum width value of approximately 0.7 s/ml is reached (Fig. 3B, inset), the suspension decreases in width, becoming mainly unimodal when small liposomes are obtained.

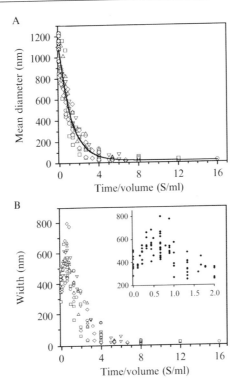

FIG. 3. (A) Time evolution of the mean diameter of the liposomes obtained in the recirculation mode of the Microfluidizer 110S as a function of the time-to-volume ratio. Volume of samples processed is 15 ml (O), 30 ml (□), 45 ml (△), 60 ml (▽), and 90 ml (◇). (B) Variation of the width of the size distribution of the liposomes obtained with the Microfluidizer 110S as a function of the time-to-volume ratio. Symbols indicate the same sample volume as shown in (A). *Inset:* Independent of the sample volume (●), at low time-to-volume ratio values, the width undergoes an increase as a consequence of the mixing of small quantities of processed sample with the large sample volume contained in the reservoir.

Effect of Temperature on Liposome Homogenization

Principle

The thermotropic properties of phospholipids determine membrane characteristics such as, for example, permeability[17] and stability.[19,20] Because of membrane rigidity below the phase transition temperature

[19] M. Wong, F. H. Anthony, T. W. Tillack, and T. E. Thompson, *Biochemistry* **21,** 4126 (1982).
[20] I. M. Hafez, S. Ansell, and P. R. Cullis, *Biophys. J.* **79,** 1438 (2000).

(T_c), some methods of liposome preparation must be performed at a temperature higher than the bilayer T_c, otherwise devices employed become blocked because, normally, the sample cannot circulate. Although in the case of high-pressure homogenizers this condition need not be accomplished when working at moderate phospholipid concentrations and high pressures, temperature is an important factor affecting vesicle size, depending on the membrane T_c.

Method

Liposome suspensions ($n = 4$) with unsaturated phospholipids at a concentration of 10 mg/ml are obtained by pouring and mixing Emulmetik 930 with water at 55° for 1 h. Liposomes made from saturated phospholipids ($n = 3$) at the previous concentration are obtained from Emulmetik 950, pouring and mixing with water for 1 h at 60°. All samples are homogenized for 1 cycle at 4 atm of inlet pressure.

The processing temperature is controlled by means of three procedures: first, the Microfluidizer coil is immersed in a water bath at the selected temperature. Second, before processing the sample, water at the chosen temperature is recirculated in the Microfluidizer in order to preheat the circuit. After this, and in order to avoid sample dilution, the water contained in the reservoir and within the pipes is taken out (e.g., by suction with a plastic pipette). Third, the sample is heated to the required temperature and placed in the reservoir.

The processed samples are analyzed with the UPA 150. As reference, three replicates of 18 ml of each type of liposome suspension are processed extensively in the recirculation mode at 23° and 4 atm of inlet pressure for 10 min.

The phase transition temperature of suspensions obtained by stirring is studied by differential scanning calorimetry (DSC) with an MC2 microcalorimeter (Microcal, Northampton, MA). Before being placed in the cell, sample aliquots are diluted with water at a final phospholipid concentration of 4 mg/ml. The scan rate is 90°/h and temperature ranges from 20 to 80°.

Results

DSC measurements of liposomes obtained from saturated phospholipids ($n = 3$) by stirring show a pretransition peak centered at 47.0 ± 0.4°, a main transition temperature at 52.9 ± 0.1°, and a small shoulder at 58.6 ± 0.3°. The last transition may reflect size inhomogeneity in the vesicle population,[21] as confirmed by size analysis with the UPA 150 (size

[21] R. L. Biltonen and D. Lichtenberg, *Chem. Phys. Lipids* **64,** 129 (1993).

distribution is bigger than 1000 nm and a part is higher than 6.5 μm, the maximum range of device analysis). Liposomes from unsaturated phospholipids show no phase transitions ($n = 3$).

It should be pointed out that homogenization causes a pressure-dependent increase in the temperature of the sample processed. In the case of 1 cycle of homogenization and between 0 and 4 atm of inlet pressure, the temperature increment is $2.12°$/atm times the inlet pressure ($r^2 = 0.9933$). Thus, considering that phospholipid-specific heat during phase transitions has no effect on suspension-specific heat temperature because of low sample concentration, and that water-specific heat is practically constant between 23 and $73°$ (changing from 4.1804 to 4.1972 J $g^{-1} °C^{-1}$, respectively), the temperature of the samples immediately after 1 cycle of homogenization at 4 atm is about $8°$ higher than that of the bath.

The effect of the (bath) temperature on the mean diameter of liposomes made from unsaturated phospholipids is shown in Fig. 4. When samples are processed for 1 cycle at 4 atm, there is no variation of mean diameter (or of width) on temperature increase. In the case of samples obtained by recirculation at 4 atm and at room temperature (Fig. 4, open circle), liposome mean diameter is significantly lower than that of the previous samples. These phenomena indicate that the constant liposome size achieved at 4 atm and 1 cycle in the studied temperature range is caused only by the

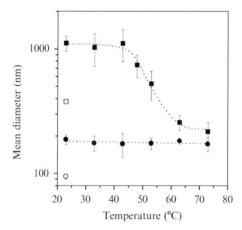

Fig. 4. Variation of liposome mean diameter with homogenization temperature. Circles: Samples obtained from unsaturated soybean phospholipids. Squares: Samples obtained from soybean hydrogenated phospholipids with a main transition temperature of $52.9°$. Solid symbols: Samples processed for 1 cycle at 4 atm of inlet pressure at the indicated temperature. Open symbols: Samples of 18 ml processed for 10 min in the recirculation mode at room temperature at 4 atm of inlet pressure.

fact that the temperature has no effect on the vesicles, because, for example, at this pressure and at room temperature liposomes can be efficiently downsized by the homogenizer if the sample is recirculated.

Unlike previous suspensions, the size distribution of liposomes from saturated phospholipids obtained by homogenization for 1 cycle at 4 atm shows temperature dependence. From room temperature to approximately 45°, liposome mean diameter is close to 1000 nm. If the temperature increases, size distribution splits into two populations: one initially centered at the previous diameter, and another at approximately 65 nm. The higher the temperature the bigger the population of small-diameter liposomes. Concomitantly, a decrease in the size of the large-diameter liposomes is observed. These facts cause a progressive diminution of the measured mean diameter. After this vesicle size decrement inside the phase transition temperature range, the size distribution of liposomes becomes constant for any temperature higher than about 60°. In these cases, the small-diameter population of liposomes is approximately 33%. Results reflect the effect of the bilayer phase state on the size reduction capability of the Microfluidizer. As reference, samples processed in the recirculation mode for 10 min at 4 atm at 23° (open square) show that, at this inlet pressure, liposome size can be controlled by homogenization at a temperature below the T_c. Under these conditions, a monomodal distribution of liposomes is obtained.

Effect of Ethanol and Phospholipid Concentration on Liposome Homogenization

Principle

As a consequence of the method of preparation or because biological implications are being studied, some liposome suspensions contain ethanol in the bulk medium. Some standard procedures for the production of liposomes, such as injection methods,[17] the Pro-Lipo-S system,[9] or the preparation of giant vesicles,[22] involve the presence of ethanol in different steps of the processes and its concentration affects the properties of the vesicles obtained. Phospholipid concentration has also shown its influence on the diameter of those liposomes obtained by the ethanol injection method[23] and by high-pressure homogenization.[6] Likewise, liposomes are employed as membrane models to investigate the effect of ethanol concentration on

[22] P. L. Ahl, L. Chen, W. R. Perkins, S. R. Minchey, L. T. Boni, T. F. Taraschi, and A. S. Janoff, *Biochim. Biophys. Acta* **1195**, 237 (1994).
[23] M. Pons, M. Foradada, and J. Estelrich, *Int. J. Pharm.* **95**, 51 (1993).

TABLE IV
EFFECT OF ETHANOL ON LIPOSOME SIZE DISTRIBUTION[a]

Equation no.	Response	R^2	F_{calc} ($F_{crit} = 4.21$)	Standard error of estimate
Eq. (9)	Mean diameter (nm) $= 288 - 118E^*$	0.982	86	102
Eq. (10)	Width (nm) $= 276 - 147E^*$	0.983	52	164

[a] Fitted empirical equations of mean diameter and width and their statistical liposome suspensions results obtained at different ethanol concentrations. The coded value of the ethanol concentration (E) is $E^* = (E - 80)/40$.

the phospholipid bilayer: membrane interdigitation,[24,25] phospholipid dehydration,[26] and ethanol association with membranes[27] have been reported.

Method

Liposome suspensions at a phospholipid concentration of 160 mg/g (this unit is used instead of mg/ml, because water–ethanol solutions are not ideal) are obtained from Emulmetik 930 by pouring and mixing with water for 1 h at 55°. After cooling the suspensions to room temperature, the necessary amounts of ethanol are poured into sample aliquots in order to obtain the desired concentration of alcohol and phospholipid. In the case of ethanol, concentration ranges from 0 to 160 mg/g, and in the case of phospholipids, it ranges from 0 to 40 mg/g. In all cases, the chosen relative concentration of the components allows liposome formation.[28] After processing the aliquots with the Microfluidizer for 2 cycles at 2.4 atm of inlet pressure, the obtained suspensions are analyzed with the UPA 150.

Results

Equations (9) and (10), shown in Table IV, are obtained by fitting the experimental mean diameter and width. Both equations have a significant factor correlation. The results show that, in the studied range, the only significant factor determining particle size is ethanol concentration.

[24] L. Löbbecke and G. Cevc, *Biochim. Biophys. Acta* **1237**, 59 (1995).
[25] H. Komatsu and S. Okada, *Biochim. Biophys. Acta* **1235**, 270 (1995).
[26] J.-S. Chiou, C.-C. Kuo, S. H. Lin, H. Kamaya, and I. Ueda, *Alcohol* **8**, 143 (1991).
[27] C. Trandrum, P. Westh, K. Jørgensen, and O. G. Mouritsen, *Biochim. Biophys. Acta* **1420**, 179 (1999).
[28] S. Perret, M. Golding, and W. P. Williams, *J. Pharm. Pharmacol.* **43**, 154 (1991).

Therefore, under the fixed operating conditions of the homogenizer, the liposome diameter diminishes with the concentration of the bulk medium ethanol and, at the same time, the vesicle size distribution becomes narrower.

Preparation of Proteoliposomes Containing Bacteriorhodopsin by High-Pressure Homogenization

Principle

Bacteriorhodopsin is a membrane protein with light-driven proton pump activity found in the purple membrane of *Halobacterium salinarum*. Bacteriorhodopsin is the simplest proton pump that, on *in vivo* absorption of light, cause a pH decrease of the outside cell medium. There are certain methods for reconstituting this protein in lipid vesicles (proteoliposomes). As bacteriorhodopsin, either as detergent-solubilized protein or as purple membrane sheets, exhibits spontaneous incorporation into large preformed vesicles,[29] the simplest procedure for obtaining proteoliposomes involves mixing of protein and liposome suspensions. There are other techniques to prepare proteoliposomes, such as sonication, French press, or reconstitution processes that involve the use of detergents.[30,31] The light-driven net flux of proton transport across the membranes of the proteoliposomes is modulated by several factors including bulk medium pH, and the composition and size of the vesicles, because they affect the preferred orientation of bacteriorhodopsin in the bilayers.[32,33] This net flux of protons can be specified by measuring the pH changes in the bulk medium.

Method

A dry film of egg yolk phosphatidylcholine (150 mg) is obtained in a round-bottom flask by rotatory evaporation. Twenty-five milliliters of a purple membrane water dispersion with a bacteriorhodopsin concentration of 0.12 mg/ml in 0.15 M KCl, pH 7, is added to the film and vortexed for 10 min in order to obtain large multilayered proteoliposomes (MLPs) caused by mechanical dispersion. A 5-ml aliquot is separated. After eliminating residual water inside the circuit of the Microfluidizer and cleaning it

[29] A. W. Scotto and M. E. Gompper, *Biochemistry* **29**, 7244 (1990).
[30] P. W. M. van Dijck and K. Van Dam, *Methods Enzymol.* **88**, 17 (1982).
[31] J. Cladera, J. L. Rigaud, J. Villaverde, and M. Duñach, *Eur. J. Biochem.* **243**, 798 (1997).
[32] M. Happe, R. M. Teather, P. Overath, A. Knobling, and D. Oesterhelt, *Biochim. Biophys. Acta* **465**, 415 (1977).
[33] K.-S. Huang, H. Bayley, and H. G. Khorana, *Proc. Natl. Acad. Sci. USA* **77**, 323 (1980).

with a small volume of the sample, the MLP suspension is processed at 23°
for 1 cycle at 4 atm of inlet pressure. An aliquot of this suspension is taken.
Finally, 15 ml of the previous suspension obtained by homogenization is
now processed in the recirculation mode at 4 atm for 6 min at 23°, in order
to achieve the minimum vesicle diameter.

Before pH measurements, all proteoliposome suspensions are kept in
darkness for at least 30 min. This procedure allows to eliminate any pH
gradient between the internal and external medium of the vesicles caused
by ambient light. All pH measurements are undertaken in dim red light.
A 2.5-ml volume of the proteoliposome suspension is transferred to a
stirred 3-ml cuvette placed in a thermostatted bath at 23°. A glass electrode
(Crison 52–08) connected to a pH meter (713 pH meter; Metrohm,
Herisau, Switzerland) is placed into the suspension. Before illuminating
the sample, about 3 min is allowed to elapse in order to permit pH stabilization. Illumination is provided by a light generator (250 PRN; Seom,
Barcelona, Spain) equipped with an incandescent halogen lamp of
250 W. The light produced is focused on the cuvette by means of a fiberoptic cable, and a yellow cutoff filter is placed between the cuvette and the
light output. The resulting light intensity on the sample is about 80×10^3 lx. Two light/darkness cycles of 3 and 7 min, respectively, are carried
out on all samples.

Results

The vesicle size distributions of the proteoliposome suspensions are
shown in Fig. 5. In the case of MLPs, part of the vesicle population has a
diameter beyond the upper range of the UPA 150 analysis (6.5 μm),
resulting in a mean diameter of 2059 ± 329 nm ($n = 5$) and a width of
810 ± 355 nm. Samples processed for 1 cycle at 4 atm have a highly spread
size distribution that includes vesicles with a diameter on the order of
60 nm to vesicles of more than 1 μm. For these suspensions, the mean
diameter is 491 ± 76 nm ($n = 4$) and the width is 540 ± 64 nm. Recirculated
samples have the smallest size distribution, with 71.4 ± 3 nm for mean
diameter ($n = 5$) and 41.6 ± 2.4 nm for mean width. Figure 6 shows the
pH increments for the different proteoliposome suspensions obtained. As
the two light/darkness cycles applied to the samples show no significant differences, results are calculated from their means. The number of independent preparations is the same as in the case of vesicle size analyses. On
illumination, all the proteoliposome suspensions produce an alkalinization
of the external bulk medium, indicating a preferred (but not necessarily a
unique) protein orientation in the bilayers, causing a net proton transloca-
tion from the external to the internal vesicle space. After illumination, a

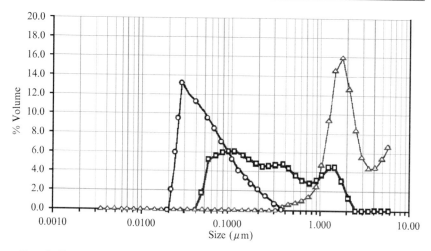

FIG. 5. Size distribution of various suspensions of bacteriorhodopsin proteoliposomes obtained by mechanical dispersion (△), by processing the previous suspension for 1 cycle at 4 atm with the Microfluidizer 110S (□), or by 6 min of recirculation in the homogenizer at 4 atm (○).

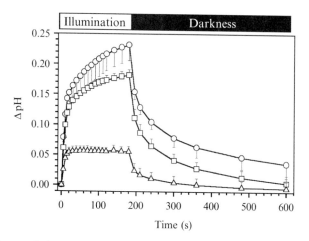

FIG. 6. Time evolution of the bulk medium pH of different proteoliposome suspensions under changing illumination conditions. The vesicle size distribution is the same as that indicated in Fig. 5. In all cases, the EPC concentration is 6 mg/ml, and the EPC-to-bacteriorhodopsin ratio is 50:1 (w/w). Results are expressed as the mean (symbols) and as positive or negative standard deviation (error bars).

pH decay is observed in all cases, being caused by passive proton diffusion from the internal aqueous space of the vesicles to the external bulk medium. MLP suspensions show a low pH increase, reaching the maximum pH value in approximately 20 s. No pH changes are detected on illumination for any time greater than 20 s. It is worth mentioning that, with this type of vesicle, when experiments are carried out that illuminate the cuvette by using a second fiberoptic cable placed opposite the first, the same results are obtained. Consequently, the low pH changes observed with MLP suspensions are not caused by the high light scattering of the samples that could act as a filter, decreasing the light across the cuvette. Compared with the previous vesicles, all homogenized proteoliposomes show a high pH increment, which is maximum in the case of recirculated vesicles. In both cases, the stationary state is not reached during the 3-min illumination period.

Acknowledgments

Purple membrane was kindly provided by the bacteriorhodopsin group directed by Dr. Esteve Padrós. We are also grateful to Ms. Yolanda Moreno and Ms. Africa Pedraza for technical assistance.

[5] Liposome Preparation by Detergent Removal

By ROLF SCHUBERT

Introduction

Detergents can be defined as the particular subgroup of surfactants that are able to solubilize lipid membranes.[1] Sufficient amounts of detergents lead to the reorganization of lipid bilayers to form smaller, soluble detergent–lipid aggregates of various shapes, which are called mixed micelles (MMs). The reverse way, that is, when the amount of detergents in MMs is reduced, leads to a successive enlargement of the MMs. At a critical detergent-to-lipid ratio membrane bilayers are formed, which spontaneously vesiculate to form liposomes. At distinct intermediate phases of detergent–membrane lipid aggregation, membrane proteins can be reconstituted into the membrane bilayers.

[1] A. Helenius and K. Simons, *Biochim. Biophys. Acta* **415**, 29 (1975).

Detergent removal for liposome preparation is superior over other methods, when essentially unilamellar liposomes are needed. This is important for their use as model membranes to study the diffusion rate of compounds through bilayers, or to quantify the transmembrane transport of compounds via reconstituted membrane proteins.[2,3] In this context a spherical shape of the liposomes is also desired, which is better realized after detergent removal than by mechanical procedures such as extrusion.[4] The production of liposomes of tailored and homogeneous sizes can be carried out more easily by detergent removal. Furthermore, when sterile liposome dispersions are required, for example, for parenteral application by infusion or injection, in principle even liposomes larger than 200 nm in diameter can be used, because the initial mixed micelle solution can be purified from microorganisms by passage through a sterile filter with pores of 0.2 μm or smaller. Finally, some methods for detergent removal offer the possibility of concentrating liposome dispersions in a second step without changing the preparation equipment.

The drawbacks of detergent removal are obvious, when excipients additional to the membrane lipids should be avoided because of increasing preparation costs or unwanted residual impurities. However, these problems can be overcome by the use of inexpensive, pure, and nontoxic detergents and there are powerful methods available to reduce the residual detergent to a level below the detection limit. Detrimental interactions of the detergent with liposomally entrapped or membrane-incorporated contents must also be considered. Detergents may lead to denaturation of biomacromolecules or may also form aggregates with smaller compounds. At least, the liposomal encapsulation efficiency of compounds can be satisfactory only when the chosen method is able to deplete the detergent effectively without removing the compound of interest.

The first part of this chapter focuses on the critical intermediate steps between solubilized membrane lipids and detergent-free bilayers, knowledge of which is necessary to optimize the preparation procedures and that can be different for various detergents. The second part of this chapter summarizes the procedures and devices for detergent removal for liposome preparation or modification, such as membrane protein reconstitution.

[2] J. V. Møller, M. le Maire, and J. P. Andersen, in "Progress in Protein–Lipid Interactions" (A. Watts, ed.), Vol. 2, p. 147. Elsevier Science, Amsterdam, 1986.
[3] F. Cornelius, Biochim. Biophys. Acta 1071, 19 (1991).
[4] R. Schubert, H. Wolburg, K.-H. Schmidt, and H. J. Roth, Chem. Phys. Lipids 58, 121 (1991).

Suitability of Detergents

Major criteria for the choice of a detergent include the desired final size of the liposomes, the rate at which they can be removed from lipid–detergent mixtures, and their required concentration to form mixed micelles. Another important aspect is the intended application for the produced liposomes. For parenteral use, detergents not tested for toxicity are excluded. For this application, bile salts are most suitable. Even under nonpathogenic conditions bile salts occur in the blood at concentrations of about 8 μM and even up to 200 μM are tolerated in cholestatic diseases.[5] Trihydroxy bile salts, in particular, such as glycocholate (GC), are nontoxic excipients, as shown for intravenously administered mixed micelles in dogs at daily doses of 20 mg of GC per kilogram body weight.[6]

Critical Micellar Concentration of Detergents

When dilute bilayer membranes are mixed with detergents, essential structural alterations occur at a detergent concentration close to the critical micellar concentration (CMC), which is therefore a crucial value for the use of detergents. At about this critical concentration, or rather in a narrow concentration range around it, not only aggregation of detergent molecules to pure detergent micelles occurs, but also solubilization of membranes to MMs. In Table I, the CMC values of detergents used commonly for liposome preparation are listed together with some aqueous media that are often employed for liposome preparation. As shown in Table I, the CMC depends not only on the detergent species, but also on other parameters such as ionic strength or pH. This is especially the case for ionic detergents, for which the degree of protonation determines the lipophilicity of the molecules and, consequently, their aggregation behavior.

Furthermore, CMC values given in the literature also depend on the methods of their measurement, which are numerous.[7] A simple method for CMC determination in small volumes is the use of fluorescent probes. The fluorescence intensity increases when the probe molecule partitions into a lipophilic environment such as a micelle, and therefore the onset

---

[5] W. F. Balistreri, *in* "Falk Symposium 80: Bile Acids in Gastroenterology" (A. F. Hofmann, G. Paumgartner, and A. Stiehl, eds.), p. 333. Kluwer Academic Publishers, Dordrecht, Amsterdam, 1995.

[6] K. Teelmann, B. Schläppi, M. Schüpbach, and A. Kistler, *Arzneimittelforschung* **34,** 1517 (1984).

[7] K. J. Mysels and P. Mukerjee, "Critical Micelle Concentrations of Surfactant Systems." Natl. Std. Ref. Data Ser. NSRDS-NBS 36. National Bureau of Standards, Washington, D.C., 1971.

TABLE I
CRITICAL MICELLAR CONCENTRATION OF DETERGENTS FOR MEMBRANE RECONSTITUTION

Detergent	Abbreviation	CMC (mM)	Remarks	Refs.
Bile Salts				
Sodium cholate	C	14.6	Water, DPH	a
		9.75	100 mM NaCl, DPH	a
		7.6	PBS, DPH	b
Sodium taurocholate	TC	10–15	Various	c
		7.0	PBS, DPH	b
Sodium glycocholate	GC	3.4–13	Various	d, e
		7.0	PBS, DPH	b
Sodium deoxycholate	DC	3–7	Various	c, f, g
		2.0	PBS, DPH	b
Sodium taurodeoxycholate	TDC	1.3–6	Various	c, g
		1.5	PBS, DPH	b
Sodium glycodeoxycholate	GDC	1.5	PBS, DPH	b
Sodium chenodeoxycholate	CDC	2.6	PBS, DPH	b
Sodium glycochenodeoxycholate	GCDC	2.05	PBS, DPH	b
Sodium taurochenodeoxycholate	TCDC	1.5	PBS, DPH	b
Other Ionic Detergents				
Sodium dodecyl sulfate	SDS	8.0	Water, DPH	a
		1.4	100 mM NaCl, DPH	g
3-[(3-Cholamido-propyl)-dimethyl-ammonio]-1-propanesulfonate	CHAPS	7.4	Water	a
Nonionic Detergents				
Octylphenolpoly-(ethyleneglycolether)$_{10}$ = Triton X-100	tert-p-C$_8\Phi$E$_{9,6}$	0.24–0.34	Water	a, c, h
		0.29	100 mM NaCl	a
Tetraethylene glycol monooctyl ether	C$_8$E$_4$	0.4	HEPES, DPH	i
Octaethylene glycol monododecyl ether	C$_{12}$E$_8$	0.087–0.11	—	g, h
n-Octyl β-D-glucopyranoside	OG	24.5	Water	h
		23.4	100 mM NaCl	a
Dodecyl maltoside	DDM	0.17	—	g

Abbreviations: PBS: 10 mM phosphate, 150 mM NaCl, pH 7.4; HEPES: 10 mM HEPES, 150 mM NaCl, pH 7.4; DPH: 1,6-diphenyl-1,3,5-hexatriene as fluorescence probe (see text).
[a] See Ref. 8.
[b] See Ref. 33.
[c] See Ref. 1.
[d] A. F. Hofmann, *Biochem. J.* **89,** 57 (1963).
[e] A. Roda, A. F. Hofmann, and K. J. Mysels, *J. Biol. Chem.* **258,** 6362 (1983).
[f] See Ref. 9.
[g] See Ref. 2.
[h] A. Helenius, D. R. McCaslin, E. Fries, and C. Tanford, *Methods Enzymol.* **56,** 734 (1979).
[i] M. Abd el Motaleb, Y.-K. Kim, and R. Schubert, unpublished data (1999).

of detergent aggregation on increasing concentration can be detected easily. As an example, for both ionic and nonionic detergents an uncharged probe such as 1,6-diphenyl-1,3,5-hexatriene (DPH) is suitable.[8]

Detergent–Lipid Aggregates

Reviews on the structural changes from MMs to vesicles or on the solubilization of vesicles to MMs show our increasing understanding of detergents, especially octylglucoside and bile salts.[9–16] In the following overview, novel findings are considered with respect to the dynamics of detergent removal. In particular, the use of bile salts is discussed. However, the general findings are probably also valid for other detergents.

Mixed Micelles

Liposome preparation by detergent removal starts with mixed micelles of detergents and membrane-forming lipids such as phospholipids, for example, lecithin, or their mixtures with cholesterol. Particular membrane lipids for the functionalization of the resulting liposomes can be added.

Distinct phases of aggregates of nonionic detergents and phospholipids are reviewed and illustrated by Ollivon et al.[16] For bile salts the aggregation behavior is similar and the stepwise shape transitions from MMs to detergent-free liposomes are pictured in Fig. 1. Even at high concentrations of bile salts, the internal detergent–phospholipid ratio in the MMs does not exceed 8:1 to 10:1 (mol/mol)[17] (one exception is ursodeoxycholate). The corresponding MMs have a size of about 10 kDa, which points to aggregates of a pair of lecithin molecules facing each other with their hydrocarbon tails and being surrounded by 16–20 bile salt molecules[18] (see Fig. 1A).

---

[8] A. Chattopadhyay and E. London, *Anal. Biochem.* **139**, 408 (1984).

[9] D. Lichtenberg, R. J. Robson, and E. A. Dennis, *Biochim. Biophys. Acta* **737**, 285 (1983).

[10] D. Lichtenberg, *Biochim. Biophys. Acta* **821**, 470 (1985).

[11] S. Almog, B. J. Litman, W. Wimley, J. Cohen, E. J. Wachtel, Y. Barenholz, A. Ben-Shaul, and D. Lichtenberg, *Biochemistry* **29**, 4582 (1990).

[12] A. Walter, *in* "Biomembrane Structure and Function: The State of the Art" (P. B. Gaber and K. R. K. Easwaran, eds.), p. 21. Adenine Press, Schenectady, NY, 1992.

[13] J. Lasch and R. Schubert, *in* "Liposome Technology" (G. Gregoriadis, ed.), 2nd Ed., Vol. II, p. 233. CRC Press, Boca Raton, FL, 1993.

[14] J. Lasch, *Biochim. Biophys. Acta* **1241**, 269 (1995).

[15] D. Lichtenberg, *in* "Handbook of Nonmedical Applications of Liposomes" (Y. Barenholz and D. D. Lasic, eds.), p. 199. CRC Press, Boca Raton, FL, 1996.

[16] M. Ollivon, S. Lesieur, C. Grabielle-Madelmont, and M. Paternostre, *Biochim. Biophys. Acta* **1508**, 34 (2000).

[17] R. Schubert and K. H. Schmidt, *Biochemistry* **27**, 8787 (1988).

[18] R. Schubert, *Proc. Molekularbiol. Biochem. Entw. Liga* **4**, 1 (1989).

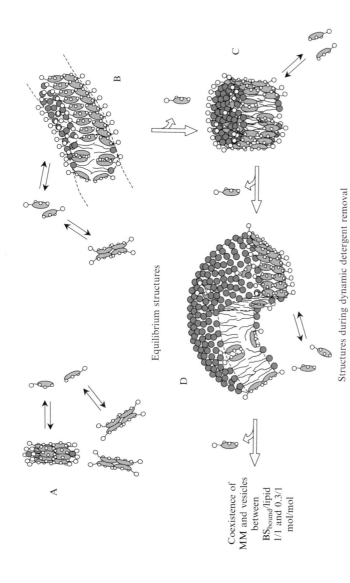

Fig. 1. Structures of mixed bile salt–phospholipid micelles. Under equilibrium conditions (A and B) rods and wormlike mixed micelles (MMs) are found, which coexist with bile salt (BS) monomers and pure BS micelles. (A) At high bile salt concentration the smallest MMs are formed by 8–10 BS molecules per lipid. (B) Suggested structure of flexible wormlike MMs with more than two BS molecules per lipid and with the same thickness as a bilayer. (C) Disklike MMs are formed under dynamic conditions during detergent removal. Defects in the "ribbon" after BS desorption lead to lateral fusion and size increase of MMs, which favors curved bilayer fragments (D, sectioned) and their vesiculation.

At a lower bile salt concentration the equilibrium structures of MMs are most probably rods or flexible polymer-like MMs, the supramolecular arrangement of which is still to be discussed. The polymer-like or "worm-like" MMs have been shown by nuclear magnetic resonance (NMR),[19] light scattering,[20,21] small-angle neutron scattering,[22] freeze–fracture,[23] and cryoelectron microscopy.[24,25] The diameter of the rods is the same as of membrane bilayers. Therefore, an early concept of diskoid[26,27] or ellipsoid[28] MMs had to be corrected, even though a transition from disks rather than from rodlike structures to the lamellar phase during liposome formation can be imagined more easily. However, it has been shown by time-resolved light and neutron scattering[29] for micellar mixtures of egg lecithin (egg phosphatidylcholine, EPC) and taurochenodeoxycholate (TCDC) that a transition from rod to disk MMs occurs when the condition switches from equilibrium to a dynamic situation during detergent removal. A suggested structure of the polymer-like MMs, which can transform easily to other MM structures, is pictured in Fig. 1B. This structure takes into account the diameter of the polymer-like MMs, which was found to be 5 nm,[25] also suggesting a bilayer-like arrangement. As shown in Fig. 1, the MM phases, at least in equilibrium, coexist with pure bile salt micelles[30] and with bile salt monomers. Until now, this coexistence has not yet been shown for nonionic detergents. For measuring the partition of detergent into mixed micelles or mixed membranes (see later) the lipid concentration in the mixture must be considered. For this purpose Lichtenberg et al.[9] introduced the effective detergent ratio

[19] J. Ulmius, G. Lindblom, H. Wennerström, L. B.-Å. Johansson, K. Fontell, O. Södermann, and G. Arvidson, *Biochemistry* **21**, 1553 (1982).

[20] D. E. Cohen, R. A. Chamberlin, G. M. Thurston, G. B. Benedek, and M. C. Carey, *in* "Falk Symposium 58: Bile Acids as Therapeutic Agents—from Basic Science to Clinical Practice" (G. Paumgartner, A. Stiehl, and W. Gerok, eds.), p. 147. Kluwer Academic Publishers, Dordrecht, The Netherlands, 1991.

[21] J. S. Pedersen, S. U. Egelhaaf, and P. Schurtenberger, *J. Phys. Chem.* **99**, 1299 (1995).

[22] R. P. Hjelm, Jr., P. Thiyagaragan, D. S. Sivia, P. Lindner, H. Alkan, and D. Schwahn, *Prog. Colloid Polym. Sci.* **81**, 225 (1990).

[23] H. Igimi and K. Murata, *J. Pharm. Dyn.* **6**, 261 (1983).

[24] A. Walter, P. K. Vinson, A. Kaplun, and Y. Talmon, *Biophys. J.* **60**, 1315 (1991).

[25] S. U. Egelhaaf, M. Müller, and P. Schurtenberger, *Langmuir* **14**, 4345 (1998).

[26] D. Small, *Gastroenterology* **52**, 607 (1967).

[27] N. A. Mazer, R. F. Kwasnik, M. C. Carey, and G. B. Benedek, *Micell. Solubil. Microemul.* **1**, 383 (1976).

[28] K. Müller, *Biochemistry* **20**, 404 (1981).

[29] S. U. Egelhaaf and P. Schurtenberger, *Phys. Rev. Lett.* **82**, 2804 (1999).

[30] P. Schurtenberger and B. Lindmann, *Biochemistry* **24**, 7161 (1985).

$$R_e = c_b/c_L \tag{1}$$

where c_b is the portion of the total detergent concentration that is bound to the membrane lipids in the mixed aggregate and c_L is the total concentration of the membrane lipid, which corresponds to the lipid concentration in the aggregates, when the monomer concentration of the lipids is negligible, as in the case of phospholipids with/without cholesterol.

Therefore, R_e represents the detergent concentration in the lipophilic MM phase and the partition coefficient is then

$$P_L = R_e/c_f = c_b/(c_L \cdot c_f) \quad (M^{-1}) \tag{2}$$

where c_f is the free detergent concentration in the aqueous phase as monomers and in pure detergent micelles.

P_L is denoted as the lipid-standardized partition coefficient and has the dimension of a reciprocal molar concentration. P_L can be recalculated[31] into a dimensionless molar partition coefficient, by

$$P_c \approx P_b = P_L \cdot 1000/MW_L \tag{3}$$

where P_c (M/M) is the molar partition coefficient according to Nernst,[32] P_b (mol kg^{-1}/mol kg^{-1}) is the molal partition coefficient, and MW_L is the mean molecular weight of the lipids in the mixture.

As an example, for pure EPC (MW $\sim$780 g mol^{-1}) the recalculation factor $1000/MW_L$ is approximately 1.28 M. For a 7:3 (mol/mol) mixture of EPC–cholesterol (MW 387 g mol^{-1}) it is approximately 1.51 M.

The total detergent concentration c_t required to result in a particular mixed detergent–lipid aggregate is

$$c_t = c_b + c_f \tag{4}$$

When mixed micelles are prepared, the monomer concentration, c_f, corresponds to the CMC value for nonionic detergents and increases to concentrations above the CMC for ionic detergents such as bile salts. Therefore there is a surplus of c_b needed over the CMC and the detergent concentration in pure detergent micelles to solubilize all membrane lipids into mixed micelles.

The lipid-standardized partition coefficients P_L^{MM} of the bile salt monomers between the MMs and the aqueous phase have been determined to be about 720 M^{-1} for cholate–EPC and 1420 M^{-1} for chenodeoxycholate–EPC.[33] For octylglucoside (OG), at a higher detergent content

[31] U. Hellwich and R. Schubert, *Biochem. Pharmacol.* **49**, 511 (1995).
[32] W. Nernst, *Zeit. Physikal. Chem.* **8**, 110 (1891).
[33] R. Schubert, Habilitation thesis. Gallensalz-Lipid-Weschelwirkungen in Liposomen und Mischmizellen. University of Tübingen, Tübingen, Germany, 1992.

P_L^{MM} to the MMs of OG and EPC is about 85 M^{-1} (calculated from data in Paternostre et al.[34]).

The formation and disintegration of detergent micelles, as well as the adsorption of detergent monomers to and desorption from the MMs, are highly dynamic processes. By decreasing the detergent monomer concentration by dilution or by monomer removal, detergent molecules are released from the MMs to follow the equilibrium binding and MMs are then forced to increase their size by lateral fusion. When the bile salt concentration is further decreased, a curvature of the mixed disk bilayers is induced (Fig. 1C). This can be explained by the concept of counteraction of the elastic forces[35] of the bilayer fragments and an edge tension induced by the detergent molecules surrounding the preformed bilayer in the diskoid MMs like a ribbon.[36]

Coexistence of Mixed Micelles and Vesicles

Below a critical detergent–lipid ratio R_e^c in the mixed micelles the curved bilayer sheets close to form vesicles. In a range between R_e^c and R_e^s (onset of membrane solubilization), vesicles and MMs coexist. The values for R_e^c or R_e^s were found to be 2.0 or 0.3 for nearly all bile salts and unsaturated phospholipids.[17] Whereas they were found to be lower, 0.3 and 0.15, respectively, for bile salts and saturated dipalmitoylphosphatidylcholine (DPPC).[37] For the nonionic detergent OG the values were determined to be approximately 2.7 or 1.7, respectively.[34] The kind of interaction of coexisting MMs and closed vesicles under dynamic conditions is still unclear. Therefore vesicle size may also be influenced by these interactions; however, it is mostly determined by the species of the detergent, which stabilizes the ribbon and fluidizes the bilayer fragment by inserting into it as a monomer or, in the case of bile salts, also as a dimer[17,27] (see Fig. 1). The membrane lipid composition also determines the elastic properties and the possible membrane curvature.

Detergent-Rich Mixed Vesicles

Below R_e^c the closed vesicles only coexist with detergent monomers. This phase of detergent–lipid mixed vesicles (MVs) is characterized by detergent-induced membrane defects (see Fig. 2, left), leading to a high

---

[34] M. Paternostre, O. Meyer, C. Grabielle-Madelmont, S. Lesieur, M. Ghanam, and M. Ollivon, *Biophys. J.* **69,** 2476 (1995).

[35] W. Helfrich, *Phys. Lett.* **50A,** 193 (1974).

[36] P. Fromherz, *Chem. Phys. Lett.* **94,** 259 (1983).

[37] C. H. Spink, V. Lieto, E. Mereand, and C. Pruden, *Biochemistry* **30,** 5104 (1991).

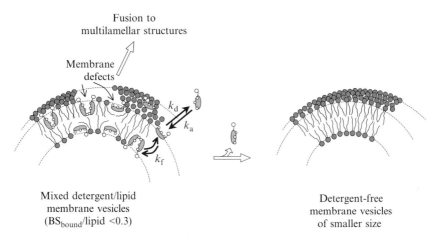

Fusion to
multilamellar structures

Membrane
defects

k_d

k_a

k_f

Mixed detergent/lipid
membrane vesicles
(BS_{bound}/lipid <0.3)

Detergent-free
membrane vesicles
of smaller size

FIG. 2. Detergent depletion from closed vesicles. *Left:* Membrane defects are formed by inserted BS monomers or dimers in detergent-rich vesicles. Defects are the reason for high membrane permeability and can result in fusion to multilamellar vesicles. Depletion of detergent molecules from the vesicles is controlled by the rate constants k_d of desorption, k_a of adsorption, and k_f of the detergent flip-flop. *Right:* Detergent-free vesicles are smaller because of the loss of membrane-inserted detergents.

probability of membrane fusion and a subsequent increase in vesicle size[38] and lamellarity.[39,40] The defects also result in a high membrane permeability. This fact should be kept in mind when compounds are to be encapsulated. To achieve detergent-depleted vesicles the stage of detergent-rich membranes should therefore be passed quickly. Membrane fluidization by the detergents is also the reason for a faster flip-flop of the membrane lipids. Membrane tensions induced during the vesiculation step by condensing the inner monolayer and stretching the outer monolayer diminish faster than after mechanical preparation procedures. This is obviously the reason for the pronounced spherical vesicles found in cryoelectron microscopy pictures compared with flatter structures after extrusion.[4]

Partition coefficients P_L^{MV} of detergents to MVs have been quantified for numerous detergents. In the case of bile salts,[33,41] OG,[34] and probably most detergents, the partition coefficient P_L^{MV} in detergent-rich MVs has a constant value over a wide detergent concentration range and is always smaller than P_L^{MM} in MMs. For EPC membranes these constant P_L^{MV} values

[38] P. Schurtenberger, N. Mazer, and W. Känzig, *J. Phys. Chem.* **89,** 1042 (1985).

[39] P. K. Vinson, Y. Talmon, and A. Walter, *Biophys. J.* **56,** 669 (1989).

[40] R. Schubert, K. Beyer, H. Wolburg, and K.-H. Schmidt, *Biochemistry* **25,** 5263 (1986).

[41] J. Lasch, V. R. Berdichevski, V. P. Torchilin, R. Koelsch, and K. Kretschmer, *Anal. Biochem.* **133,** 486 (1983).

were found to be 50 M^{-1} for cholate and 250 M^{-1} for chenodeoxycholote (CDC)[14,33] and 18 M^{-1} for OG.[34]

Detergent-Poor Mixed Vesicles and Detergent-Free Vesicles

P_L^{MV} increases continuously, when a limiting R_e^{dr} (effective detergent ratio in detergent-rich MVs) is not reached, up to a 2- to 5-fold value of the constant P_L^{MV} for different detergents. The interaction of bile salts containing a sterol backbone with membrane lipids in detergent-poor MVs is more specific than that of detergents with hydrocarbon chains such as OG. For all bile salts a specific binding site in the membrane has been found, that is, as monomers or dimers, bile salt molecules are surrounded by six lecithin molecules.[17]

To achieve detergent-free vesicles, the detergent molecules must move from the inner to the outer membrane monolayer first (see Fig. 2). Most of the detergents show a fast flip-flop in EPC membranes, with a half-life, $t_{1/2}$, of several milliseconds for CDC or deoxycholate (DC), which are lipophilic dihydroxy bile salts, and about 140 ms for cholate as the more hydrophilic trihydroxy bile salt at neutral pH. However, at pH 10, $t_{1/2}$ is about 24 h.[42] This clearly suggests that at acid–base equilibrium a part of the detergent should be uncharged to ensure complete depletion within a reasonable time. For the nonionic detergent octaethylene glycol monododecyl ether ($C_{12}E_8$), $t_{1/2}$ was found to be about 100 ms.[43] Nevertheless, a high residual amount of this detergent was found, the reason for which is not clear.[44]

The depletion of detergents from detergent-rich MVs [bile salts–lipid, 0.3:1 (mol/mol)] forming detergent-free vesicles is accompanied by a decrease in vesicle size (see Fig. 2), which is inevitable considering that one-quarter of the molecules in the membrane are lost.

The pathways from mixed micelles to detergent-depleted vesicles may differ slightly for the various detergent species, especially when bile salts or nonionic detergents such as OG are compared.[14,16] However, the pathway suggested in Figs. 1 and 2 may be helpful to yield optimum liposome preparation by detergent removal.

It should be noted that the pathway from MMs to detergent-free vesicles is not completely reversible, when detergents are added to vesicles to solubilize the membranes and to achieve MMs. The differences are pronounced, for example, in an asymmetric adsorption of added detergents such as $C_{12}E_8$[44] and bile salts[40] to the outer monolayer. Transient and large

[42] D. Cabral, D. Small, H. S. Lilly, and J. A. Hamilton, *Biochemistry* **26**, 1801 (1987).

[43] M. le Maire, J. Møller, and P. Champeil, *Biochemistry* **26**, 4803 (1987).

[44] M. Ueno, C. Tanford, and J. A. Reynolds, *Biochemistry* **23**, 3070 (1984).

membrane defects induced by added bile salts or OG were also observed. These may result from the membrane tension caused by asymmetric detergent adsorption, which leads to parts of the outer monolayer suddenly being refolded to the inner monolayer. These "pores"[40] exist only for seconds to minutes[13] and can even be used to encapsulate substances, with high encapsulation efficiency of up to 50%, into preformed vesicles (detergent-induced liposome loading, DILL[45]) or for the determination of vesicle lamellarity.[4]

Parameters Determining Vesicle Size

The ultimate vesicle size, as shown in Table II, depends on a variety of factors that influence the dynamic equilibriums between MMs, pure micelles, and detergent monomers.

Detergent Species

Detergent species show a different partition to the edges and on the surfaces of MMs, depending on their charge, shape, and lipophilicity. When using a homologous series of a detergent group, such as bile salts or alkylglucosides, their vesicles size correlates with the lipophilicity of the detergent (see Table II). This may be due to slower removal of lipophilic detergents when using the same depletion method. This, in turn, leads to a higher probability for lateral fusion and enlargement of diskoid MMs before vesiculation occurs.

Initial Ratio of Detergent to Lipid

The minimum required detergent concentration is approximately the CMC, when the lipid concentration is lower. At a higher lipid concentration the initial ratio depends on the detergent partition to the MMs. As shown previously, the structure and size of the MMs are determined by the detergent–lipid ratio in the MMs. A lower equilibrium detergent content correlates with smaller MMs. On dynamic detergent removal, fast detergent desorption probably produces transient lipophilic defects in the MMs. Lateral fusion and vesiculation then compete to reseal these defects. Starting with smaller MMs therefore results in smaller vesicles.

When using trihydroxy bile salts such as sodium cholate, the initial total molar detergent–lipid ratio should be between 2.0 and 1.2 with up to 25 mM lipid to result in clear MM solutions; for dihydroxy bile salts a ratio of only approximately 0.5 is needed. For OG, which is a hydrophilic

[45] R. Schubert, *Proc. Molekularbiol. Biochem. Entw. Liga* **5,** 73 (1990).

nonionic detergent, a total molar detergent–lipid ratio of 5:1 should normally yield a clear MM solution with 25 mM lipid.

Total Concentrations of Detergent and Lipid

As shown previously, because of the equilibrium between detergent as monomers, in pure micelles, and in MMs, increasing the lipid amount requires a nonlinear increase in the required detergent to achieve the same type of MMs. However, on detergent removal a higher concentration of MMs in the solution leads to a higher probability of fusion of MMs before vesiculation, or of interaction between coexisting vesicles and MMs. Therefore a higher starting lipid concentration results in larger vesicles.

One should carefully consider also the increased tendency of fusion of detergent-rich MMs to multilamellar structures on increasing the lipid concentration. Therefore a higher lipid concentration requires a method for faster detergent removal (see later) to result in unilamellar vesicles.

Lipid Species, Their Phase Transition Temperature, and Charge

The flexibility, charge, and structure of lipids determine the resulting vesicle size. In most cases, unsaturated natural glycerophospholipids with a cylindrical molecular shape (e.g., phosphatidylcholine and phosphatidylserine) can be used as single components or as mixtures without problems. They have phase transition temperatures below 0° and form fluid bilayers. With increasing amounts of stiff membrane lipids the successive transformation of MMs to vesicles can be disturbed. As an example, using sodium cholate as detergent, sphingomyelin (SM) from egg yolk, which has almost saturated side chains, can be mixed with EPC only up to approximately 30 mol% to yield unilamellar liposomes, when prepared at room temperature.[46] The addition of SM to EPC increases the final vesicle diameter from about 70 nm for pure lecithin to 105 nm at 30 mol% SM.

To avoid multilamellar structures, unsaturated lipids or mixtures should generally be prepared at least 10° above the main phase transition temperature T_m of the membrane lipids (reviewed in Szoka and Papahadjopoulos[47]). However, not all species of membrane lipids are suitable for detergent removal procedures. Therefore a first criterion for the lipid suitability is a clear MM solution with the chosen detergent to obtain homogeneous vesicle populations at the preparation temperature.

When cholesterol is part of the lipid mixture, only an amount less than that used in mechanical procedures can be added. In our experience, when

---

[46] P. Troschel, Ph.D. thesis. University of Tübingen, Tübingen, Germany, 1986.
[47] F. Szoka and D. Papahadjopoulos, *Annu. Rev. Biophys. Bioeng.* **9,** 467 (1980).

TABLE II
LIPOSOME SIZE AFTER DETERGENT REMOVAL[a,b,c]

A. Controlled Membrane Dialysis[a,b,c]

Detergent	Membrane lipid (mol/mol)	Initial detergent/lipid ratio (mol/mol)	Final mean vesicle diameter (nm)[d,e]	Preparation conditions	Ref.
C	EPC	2:1	40	Water, room temperature	f
C	EPC	2:1	70	PBS, room temperature	g
C	EPC–Chol (7:3)	2:1	80	PBS, 4°	g
			75	PBS, room temperature	
GC	EPC	2:1	100	PBS, 4°	h
TC			80	PBS, room temperature	
DC	EPC	0.45:1	80		h
CDC			155	PBS, room temperature	
HG	EPC	7:1	160		i
OG	EPC	5:1	80	HEPES, room temperature	f
			140	Water, room temperature	
C_8E_5	EPC	5:1	180	PBS, room temperature	j
			~200 (OLV) and ~3000 (MLV)	22°, PBS	
	EPS		75		
	EPC–EPS (9.5:0.5)		820		
	EPC–EPS (7:3)		470		
	EPC–EPS (1:1)		240		
	EPC–EPS (3:7)	5:1	140	22°, PBS	j
			120	15°, PBS	
			460	37°, PBS	

(continued)

TABLE II (*Continued*)

B. Other Methods[k]

Removal method	Detergent	Membrane lipid (mol/mol)	Initial lipid concentration (mM)	Initial detergent/lipid ratio (mol/mol)	Final mean vesicle diameter (nm)[d,e]	Preparation conditions	Ref.
Dilution	GC	EPC	65	3.3:1	100	Fast dilution, 1:15; Tris, pH 8	l
					24	Fast dilution, 1:60; Tris, pH 8	
	OG	EPC–Chol (8:2)	12.5	15:1	118	Dilution, 1:11 in 5 s; Tris, pH 7.4	m
					194	Dilution, 1:11 in 7 min; Tris, pH 7.4	
Gel chromatography	C	EPC	13	2:1	30	Sephadex G-50; Tris, pH 7.3	n
Hollow fibers (small cartridge)	C	EPC	20	4:1	104	pH 6; 10 mM Tris–sucrose	o
					90	pH 7; 10 mM Tris–sucrose	
					70	pH 8; 10 mM Tris–sucrose	
Cross-flow	C	SPC	10	1.6:1	50	MOPS	p
				8.5:1	35		
		SPC–Chol (2:1)		1.6:1	45		
		SPC		5:1	145		
Beads	OG						
	OG	EPC–PS–Chol (1:1:1)	10	10:1	240	<0.12 μmol of OG per mg XAD-2 beads	q
	OG	EPC	15	12:1	150	XAD-2	r
	$C_{12}E_8$			10:1	85		

Abbreviations: EPC, Egg phosphatidylcholine (egg lecithin); SPC, soy lecithin; Chol, cholesterol; EPS, egg phosphatidylserine; C, cholate; GC, glycocholate; TC, taurocholate (*note*: C, GC, and TC are trihydroxy bile salts); DC, deoxycholate; CDC, chenodeoxycholate (*note*: DC and CDC are dihydroxy bile salts); HG, heptylglucoside; OG, octylglucoside; C_8E_4, tetraethylene glycol monooctyl ether; $C_{12}E_8$.

octaethylene glycol monododecyl ether; PBS: 10 mM phosphate, 150 mM NaCl; Tris: 10 mM Tris, 150 mM NaCl; MOPS: 10 mM MOPS, 150 mM NaCl (all buffers between pH 7.3 and 7.4).

[a] When liposomes are prepared by fast detergent dialysis, thickness and pore size of the dialysis membranes (<10 μm thickness) influence liposome size only slightly ($\sim\pm10\%$). Fast dialysis of small volumes (~1 ml) in rotating dialysis chambers results in about 20% smaller vesicles than with larger dialysis devices.

[b] Lipid concentration in all cases was 10–20 mM. In this range and below, vesicle size is almost uninfluenced by lipid concentration.

[c] For a review on controlled detergent analysis, see H. G. Weder and D. Zumbuehl, in "Liposome Technology," (G. Gregoriadis, ed.) 1st Ed. Vol. 1, p. 79. CRC Press, Boca Raton, FL, 1983.

[d] Liposome size is in most cases measured by photon correlation spectroscopy. Data are normally intensity weighted. Deviations of the given diameter are on the order of 20% because of the use of different PCS instruments and software, variation in the composition of natural mixtures of phospholipids such as egg lecithin, variation of room temperature, and so on.

[e] Vesicles are essentially unilamellar with a narrow size distribution when no other remarks are given. Lamellarity is determined by negative staining or cryotransmission electron microscopy, or by ^{31}P NMR.

[f] I. Ohlhoff and R. Schubert, unpublished data (2001).

[g] See Ref. 33.

[h] M. Wacker, Ph.D. thesis. University of Tübingen, Tübingen, Germany, 1988.

[i] S. Wieland and R. Peschka-Süss, unpublished data (1998).

[j] E. Stocker, Ph.D. thesis. University of Tübingen, Tübingen, Germany, 1991.

[k] For a comparison of detergent analysis and bead adsorption, see J. R. Philippot.

[l] See Ref. 38.

[m] See Ref. 53.

[n] See Ref. 54.

[o] See Ref. 50.

[p] See Ref. 60.

[q] See Ref. 64.

[r] M. Ueno, Biochim. Biophys. Acta **904**, 140 (1987).

sodium cholate is used as detergent, a critical upper limit is 30 mol%, compared with about 50 mol% with extrusion or homogenization. Nonionic detergents form stable MMs with phospholipids even at lower cholesterol content.

Unlike vesicles of dipalmitoylphosphatidylcholine (DPPC, T_m 41°), homogeneous unilamellar vesicles of dimyristoylphosphatidylcholine (DMPC, T_m 23°) are practically impossible to prepare by cholate dialysis. This is true even at preparation temperatures above the T_m. The reason for this is probably an additional interaction of the bile salt with DMPC membranes also in the deeper hydrocarbon region, which does not occur with DPPC.[48] The charge of the lipid also influences vesicle size. Increasing amounts of phosphatidylserine (PS) from porcine brain in mixtures with EPC show a pronounced size decrease. This is true for the use of bile salts as well as for octylglucoside.

Preparation Temperature

Temperature influences the CMC of detergents, the equilibrium of MM solutions, and the fluidity of membranes. As pointed out above, the preparation temperature should be approximately 10° above the T_m. At the T_m itself, DMPC vesicles prepared by dilution of bile salts are three times larger than those prepared at 40°.[49] Nonionic detergents are dehydrated at increasing temperature, which makes them more lipophilic. The temperature must be below the cloud point, at which the detergent is dehydrated and forms a gel or precipitates. The cloud points of individual detergents can be found in the manufacturer's information.

Ionic Strength

Strong influences of ions on ionic detergents and lipids are to be expected.[38] In general, then, ionic strength correlates with final vesicle size. However, detergent removal from MMs consisting of nonionic detergents such as OG and uncharged EPC is also influenced by ion strength in a similar manner (see Table II).

pH of Aqueous Media

The pH value plays an important role in the transition from MMs to vesicles, when ionic detergents such as weak acids, bile salts,[50] or bases are used. Below pH 6, unconjugated bile salts such as sodium cholate

[48] T. M. Bayerl, G.-D. Werner, and E. Sackmann, *Biochim. Biophys. Acta* **984**, 214 (1989).
[49] P. Schurtenberger, R. Bertani, and W. Känzig, *J. Colloid Interface Sci.* **114**, 82 (1986).
[50] V. Rhoden and S. M. Goldin, *Biochemistry* **18**, 4173 (1979).

precipitate at higher concentration and are then useless for liposome preparation starting with MMs. At a lower pH, either nonionic detergents or bile salts with a lower pK_a value, such as glycocholate or taurocholate, may be used to dissolve higher lipid amounts. As shown previously, an increasing amount of residual bile salts may be found at a high pH, for example, pH 10, caused by low flip-flop rates of the bile salts remaining on the inner membrane monolayer.[42] Therefore, for bile salt removal the pH should be only slightly above the pK_a value.

Rate of Detergent Removal

As pointed out previously, when discussing the stepwise transition from MMs to detergent-free vesicles, fast detergent removal leads to decreased vesicle size but avoids side effects such as fusion of detergent-rich vesicles to multilamellar structures, which would be a problem at high lipid concentrations above 20 mM. In each case the rate of detergent removal should be determined in initial studies to decide about a suitable removal technique. In the case of lipophilic detergents, a fast depletion method should be chosen.

Molecules to Be Encapsulated

Drugs to be entrapped into liposomes or incorporated in liposomal membranes can interact with detergent or lipids. The interaction of surface-active drugs with membranes and surfactants was reviewed by Schreier et al.[51]

Methods for Detergent Removal

Preparation of Mixed Micelle Solutions

A mixed micelle solution can be produced in two different ways. The lipids and lipophilic substances to be incorporated into the liposomal membrane are dissolved together with the detergent in a suitable organic solvent or solvent mixture to obtain a clear solution. In most cases methanol, ethanol, or mixtures with chloroform are suitable. The solvent is then removed in a rotary evaporator by reduced pressure at a moderate temperature. Residual solvent should be removed by high vacuum for at least 1 h. The dry film is normally clear, when bile salts are used as detergent. With nonionic detergents the film may be turbid. A suitable buffer, optionally together with hydrophilic substances to be encapsulated, is added to

[51] S. Schreier, S. V. P. Malheiros, and E. de Paula, *Biochim. Biophys. Acta* **1508**, 210 (2000).

yield the desired lipid concentration and the temperature is adjusted. With OG, after adding the buffer, the dispersion may be opalescent for some seconds before clearing. Alternatively, a preformed liposome dispersion may be dissolved successively with detergent at the desired preparation temperature until a clear solution is achieved.

A main MM solution is also a prerequisite for preparing homogeneous liposomes by the first method. If there are still particles in the MM solution, the detergent content may be too low, another detergent may be required, or the lipid mixture may not be suitable for detergent removal procedures. If the particles are not substances of interest, they can be removed by filtration through filters with 0.2-μm pores.

Preparation of Liposomes

Dilution. Dilution of MM solutions has the advantage that desorption from MMs and removal of detergent monomers are not hindered by any diffusion barrier except the aqueous solvent. Therefore the depletion rate is high and the final equilibrium compositions are well defined by the dilution factor. As shown in numerous studies, dilution is therefore an excellent method by which to study the principal parameters of mixed micelle to vesicle transition using bile salts[38,52] or OG.[53]

The obvious drawback of the method is the low encapsulation efficiency or the high amount of substance needed, when it should be encapsulated at high concentration. Because of the high volume after dilution, the separation of free compound from entrapped compound is laborious.

Gel Chromatography. During the column run of a MM solution, the larger MMs are continuously separated from detergent monomers, and pure micelles and liposomes are formed on the column. This method was first reported by Brunner *et al.*[54] to produce homogeneous and unilamellar vesicles from cholate–EPC MMs. The size was determined to be about 30 nm in diameter, which indicates fast detergent removal.

The size of detergent-rich MMs is close to 10,000 Da, but is more commonly about 50,000 Da in starting MM solutions. When the exclusion size of the gel network is larger than the growing MMs, the liposomes, which form directly on the column, can be trapped in the gel beads and lipid recovery after the run may be lower than expected. In addition, the liposome dispersion is then strongly diluted on the column. Therefore the exclusion

[52] K. Son and H. Alkan, *Biochim. Biophys. Acta* **981,** 288 (1989).
[53] W. Jiskoot, T. Teerlink, E. C. Beuvery, and D. J. A. Crommelin, *Pharm. Weekblad Sci. Ed.* **8,** 259 (1986).
[54] J. Brunner, P. Skrabal, and H. Hauser, *Biochim. Biophys. Acta* **455,** 322 (1976).

size should be smaller than the initial MM size. This leads to lesser dilution of the liposomes in the void volume and removes the detergent monomers more effectively from the formed vesicles.

A main drawback of the method is that most hydrophilic substances of interest are also separated from the liposomes during preparation. To achieve encapsulation, the molecular mass of the substances must also be larger than the exclusion size; otherwise the column must be preequilibrated with the substance to be encapsulated, which then yields only a low percentage of encapsulation efficiency. A further preparative problem is that for removal of residual detergent or nonencapsulated drugs, additional column runs must be performed or other methods must be used.

Dialysis Using Flat Membranes. Detergent monomers and micelles may be separated from MMs and the forming vesicles by diffusion through dialysis membranes. To maintain the highest possible detergent concentration gradient for the diffusion, continuous flow of the dialysis buffer at the membrane is recommended. Alternatively, the volume of the buffer dialysate must be large, that is, at least 100-fold, compared with the retentate, which contains the detergent–lipid aggregates. The retentate should be mixed continuously to avoid inhomogeneous MM and/or vesicle structures. This can be done with a magnetic stirrer, by pumping the lipid-containing dispersion tangentially to the membrane [e.g., by using a Liposomat (Dianorm, Munich, Germany)] or by allowing a small air bubble to remain in the retentate in rotating dialysis chambers (e.g., Mini-Lipoprep, 1-ml retentate volume; Dianorm). To ensure effective and fast detergent removal, the thickness of the commercially available or self-made dialysis cell should be small, whereas the contact area of the solution with the membrane should be large. Normal dialysis bags are ineffective for detergent removal because of their thick membranes. This can result in the partial formation of multilamellar structures. Most suitable are highly permeable membranes with a thickness of less than 10 μm (e.g., Dianorm). The pore size (cutoff) of the membranes also determines the detergent removal rate and at least the liposome size. In most cases it can be up to 50,000 Da, but normally membranes with a cutoff of 10,000 or 5000 Da are used.

The method of controlled dialysis was first described by Milsmann *et al.*[55] Vesicle formation is relatively fast for hydrophilic detergents. Using cholate, the turbidity in the retentate is at a maximum after 2–3 h, indicating vesicle formation. With OG and some other nonionic detergents with high CMC, vesicles are formed after less than 1 h. To deplete residual

[55] M. Milsmann, R. A. Schwendener, and H.-G. Weder, *Biochim. Biophys. Acta* **512,** 147 (1978).

detergent, dialysis against detergent-free buffer should be continued for at least another 16 h (overnight). After that time, the residual detergent in EPC vesicles was found to be below the detection limit of 1 per 100 lipid molecules. However, using membrane dialysis, more lipophilic detergents such as dihydroxy bile salts, for example, chenodeoxycholate and deoxycholate, are only slowly depleted from the mixed micelle solution, and detergent-rich liposomes with still about 15% of the initial bile salt amount are formed only after 20 h.[56]

The removal by dialysis results in homogeneous unilamellar vesicles, when the concentration of fluid lipids does not exceed approximately 25–30 mM. The inner aqueous volume of homogeneous liposomes that are 70 nm in diameter is then 5% (for the calculation formula, see below), which should also be the encapsulation efficiency (EE), when hardly any substance to be encapsulated is lost from the retentate. This could be shown for [^{3}H]inulin (molecular weight, 5000) and dialysis membranes with a cutoff of 5000 Da.[57] After OG removal vesicles have a mean diameter of about 180 nm, and therefore a maximum EE of 15% under the same conditions. However, most substances of interest are smaller than the pores of the dialysis membrane and would be removed in the same way as the detergent. To achieve a reasonable drug loading, the desired concentration of the substance in the liposomes must therefore be maintained in the dialysis compartment and in the dialysis buffer until closed detergent-poor vesicles are formed.

Dialysis Using Hollow Fibers. Detergent removal using small hollow-fiber cartridges was reported by Rhoden and Goldin.[50] The contact area of the MM solution with the hollow fiber membranes is much larger than that with dialysis membranes, and therefore removal occurs even faster. An additional advantage is that large cartridges for blood dialysis are highly suitable for upscaled liposome production.[58,59] The cutoff of the fibers may be 5000 or 10,000 Da. Using flow rates of 12 ml/min in the MM compartment and 24 ml/min in the external buffer compartment, even with the smaller pore size liposome formation is almost completed in a single dialysis step.[59] After three further discontinuous dialysis runs, liposomes are detergent poor. Starting with sterile-filtered mixed micelle solutions, the cartridge makes aseptic preparation of the liposomes possible, which is essential for parenteral or ocular application.

[56] M. Wacker and R. Schubert, *Int. J. Pharm.* **162**, (1998).
[57] H. Jaroni and R. Schubert, unpublished data (1983).
[58] R. A. Schwendener, *Cancer Drug Deliv.* **3**, 123 (1986).
[59] P. Hirnle and R. Schubert, *Int. J. Pharm.* **72**, 259 (1991).

Liposome dispersions can be easily concentrated afterward, using the same device. When the dialysis buffer is substituted by a buffer containing 10% (w/v) polyethylene glycol (molecular weight, 40,000), 85% of the water of the liposome dispersion is removed within 90 min.

Filtration. The removal of detergent can be accelerated by actively separating detergent monomers from the lipid aggregates by filtration. Any pressure perpendicular to the filter surface would plug the pores with MMs or liposomes. This will not occur if the fluid is moving under low pressure tangentially to the filter surface. This is achieved by tangential or cross-flow filtration[60] [e.g., Pall Filtron (Northborough, MA), Millipore (Bedford, MA), Sartorius (Edgewood, NY)]. Removal of the detergent in this way reduces the volume of the retentate and concentrates the lipid, but the detergent concentration remains constant. The equilibrium between lipid aggregates and detergents is therefore not shifted enough to result in liposome formation. However, when the volume of the rapidly filtered-off solution is continuously substituted for detergent-free buffer, the lipid concentration remains constant and the resulting dilution of the detergent in the retentate leads to fast formation of unilamellar, homogeneous, and detergent-free liposomes.

Commercially available membrane cassettes with filter cutoffs of 10,000 or 50,000 Da, and that are designed for low protein binding, may be used. Both cutoffs lead to the same preparation time, whereas 100,000-Da pores are too large and result in a loss of lipid. The crucial parameter is the filtration area, which can be manipulated by using particular cassettes that can be combined. Increasing the area from 140 cm^2 (0.15 ft^2) to 700 cm^2 (0.75 ft^2) reduces the preparation time by a factor of about 10. At a flow rate of 250 ml/min, which induces an additional pressure of 0.4 atm at the filtration membrane, 100 ml of MM solution with 16 mM cholate and 10 mM soy lecithin yields detergent-rich liposomes (molar detergent–lipid ratio, <0.3) within 12 min. Liposomes are essentially cholate free [cholate–lecithin, 1:100 (mol/mol)] after 60 min, and are smaller (45 nm) than with membrane dialysis (70 nm). Surprisingly, the presence of cholesterol up to 33 mol% in the lecithin membrane does not increase vesicle size.

The continuous substitution of wasted filtrate with buffer is effective in the removal of detergent. Only 300 ml of filtrate is formed, yielding 100 ml of detergent-free liposomes.[61] The lipid concentration can be much higher than with the dialysis method. In buffer (10 mM HEPES, 150 mM NaCl, pH 7.4), the resulting liposomes at 50 mM lipid are then still unilamellar and somewhat larger (70 nm). At 100 mM lipid the mean size of the

[60] R. Peschka, T. Purmann, and R. Schubert, *Int. J. Pharm.* **162**, 177 (1998).
[61] R. Peschka-Süss, T. Dern, and R. Schubert, unpublished data (2000).

still homogeneous dispersion is 130 nm. An encapsulation efficiency of approximately 30% for larger molecules (molecular weight, >5000), compared with the calculated value of 47%, suggests the presence of some oligolamellar vesicles.[61]

When the membranes are preincubated in 0.1 N NaOH and then are washed with sterile buffer, cross-flow filtration can also be used for aseptic liposome production.

Concentrating the liposomes after formation and detergent depletion is easy with cross-flow filtration. Without substituting the pressed-off fluid, the dispersion can be concentrated almost to a gel without any additional excipient and without changing the preparation device. A drawback of the method may be the relatively large minimum required volume of 30 ml for aqueous dispersion. This makes the preparation suitable only for upscaled production.

Adsorption to Beads. Suitable hydrophobic and porous polymer particles such as polystyrene Bio-Beads SM-2 or SM-4 (Bio-Rad, Hercules, CA),[62,63] Amberlite XAD-2 beads (Sigma, St. Louis, MO),[44,64,65] or ion-exchange resins (e.g., cholestyramine; Bristol-Meyers Squibb, New York, NY) adsorb nonionic detergents or bile salts effectively. The adsorption capacity of hydrophobic beads is on the order of 100 mg of detergent per gram of beads or even higher. Beads are especially recommended for removing hydrophobic detergents, such as dihydroxy bile salts, $C_{12}E_8$, or Triton X-100, from mixed detergent–lipid aggregates[63] or from reconstituted membrane proteins[66] in a few hours. For OG with high CMC, the removal is finished after about 10 min.[64] Lipid is also adsorbed by the beads, but to an extent of only 1% compared with detergent. This makes bead adsorption suitable when dilution or loss of lipid is to be avoided, or for small samples, for example, in the reconstitution of membrane proteins. The beads can be used up to 10 times, but before reuse they need to be cleaned carefully with methanol and then washed thoroughly with water.

Others. Combinations of the listed methods may be advantageous when different detergents are used, that is, for protein isolation and reconstitution into membranes. In addition to the listed and most frequently used methods for detergent removal, others such as temperature and pressure jumps,[67–69] or the enzymatic cleavage of detergents and their conversion to membrane lipids, are special applications.[70]

[62] J. Philippot, S. Mutaftchiev, and J.-P. Liautard, *Biochim. Biophys. Acta* **734**, 137 (1983).
[63] J. A. Reynolds and D. R. McCaslin, *Subcell. Biochem.* **14**, 1 (1989).
[64] J. Philippot, S. Mutaftchiev, and J.-P. Liautard, *Biochim. Biophys. Acta* **821**, 79 (1985).
[65] M. Ueno, N. Tanaka, and I. Horikoshi, *J. Membrane Sci.* **41**, 269 (1989).
[66] D. Levy, A. Bluzat, M. Seignoret, and J. L. Rigaud, *Biochim. Biophys. Acta* **1025**, 179 (1990).

Protein Reconstitution. A detailed description of procedures to reconstitute membrane proteins by the use of detergents is beyond the scope of this chapter and can be found in numerous studies and reviews,[2,3,63,71–74] as well as in Chapter 4 of Volume 372.[75]

In principle, the same methods are used as for the preparation of protein-free liposomes, although detergent removal is achieved mostly with beads or by dialysis. When bile salts are used, reconstitution starts with vesicle preparation from MMs. With nonionic detergents such as OG or Triton X-100, best results are obtained when the detergent is added to preformed vesicles to achieve detergent-rich vesicles with large membrane defects (see Fig. 2), into which membrane proteins such as bacteriorhodopsin can be inserted.[76] For transport studies, the residual detergent then must be depleted effectively to minimize the fluidity of the bilayers to the values of natural membranes.

Control of Liposome Quality

Size, structure, lipid concentration, residual detergent, and stability of the liposomes—and optionally the encapsulation efficiency of substances—should be determined after preparation.

Size, Lamellarity, and Shape of Liposomes. Vesicle size is determined routinely by dynamic or static light scattering with numerous commercially available devices. However, for a more careful study of size distribution and vesicular structures, electron microscopic studies are recommended.

Vesicle lamellarity can be determined by cryoelectron microscopy (cryo-EM),[4,24] which also provides information about liposome shape, or by ^{31}P NMR using shift reagents such as Pr^{3+} or Eu^{3+}, rather than Mn^{2+}.[77,78] Both methods are restricted to liposomes smaller than

[67] M. G. Miguel, O. Eidelman, M. Ollivon, and A. Walter, *Biochemistry* **28,** 8921 (1989).

[68] A. I. Polozova, G. E. Dubachev, T. N. Simonova, and L. I. Barsukov, *FEBS Lett.* **358,** 17 (1995).

[69] P. Lesieur, M. A. Kiselev, L. I. Barsukov, and D. Lombardo, *J. Appl. Crystallogr.* **33,** 623 (2000).

[70] J. Chopineau, S. Lesieur, B. Carion-Taravella, and M. Ollivon, *Biochimie* **80,** 421 (1998).

[71] J. R. Silvius, *Annu. Rev. Biomol. Struct.* **21,** 323 (1992).

[72] D. Levy, A. Gulik, A. Bluzat, and J.-L. Rigaud, *Biochim. Biophys. Acta* **1107,** 283 (1992).

[73] J. Knol, L. Veenhoff, W.-J. Liang, P. J. F. Henderson, G. Leblanc, and B. Poolman, *J. Biol. Chem.* **271,** 15358 (1996).

[74] M. M. Parmar, K. Edwards, and T. D. Madden, *Biochim. Biophys. Acta* **1421,** 77 (1999).

[75] J.-L. Rigaud and D. Levy, *Methods Enzymol.* **372,** 4 (2003).

[76] J.-L. Rigaud, M.-T. Paternostre, and A. Bluzat, *Biochemistry* **27,** 2677 (1988).

[77] M. Fröhlich, V. Brecht, and R. Peschka-Süss, *Chem. Phys. Lipids* **109,** 103 (2001).

[78] N. Düzgüneş, J. Wilschut, K. Hong, R. Fraley, C. Perry, D. S. Friend, T. L. James, and D. Papahadjopoulos, *Biochim. Biophys. Acta* **732,** 289 (1983).

approximately 200 nm. Much larger liposomes cannot be representatively seen in the thin shock-frozen aqueous films for cryo-EM. Powder signals of large vesicles in ^{31}P NMR can hardly be analyzed for asymmetric distribution of the shift ions. Alternative methods include negative- or positive-staining transmission electron microscopy or use of the deviation of the encapsulation efficiency calculated for unilamellar vesicles from the actual values.[4]

Encapsulation Efficiency. The trapping or encapsulation efficiency of a substance of interest can be given as its liposome-associated mass per lipid mass, or as the encapsulated percentage of the initial amount. The portion of the aqueous interior phase of liposomes in relation to the total aqueous volume can be calculated by assuming the vesicles are unilamellar and spherical. Then the calculated value[4] compares with the theoretical maximum encapsulation efficiency EE for hydrophilic substances, when no drug is lost from the vesicular compartment during the preparation.

$$\text{EE}(\%) = \frac{(R - d)^3}{\left(R - \frac{d}{2}\right)^2 \cdot \left(\frac{0.09963}{c_L \cdot a} - 0.03 \cdot d\right)} \qquad (5)$$

where R is the mean hydrodynamic radius of a homogeneous vesicles population, d is the thickness (nm) of the lipid bilayer ($\sim$6 nm for EPC), a is the cross-section area (nm^2) of a single lipid in the bilayer ($\sim$0.75 nm^2 for EPC), and c_L is the total lipid concentration (M).

The calculated values make sense only up to 50%, when vesicles are densely packed. Any removal of the substance of interest from the lipid aggregate compartment during liposome preparation and/or deviation from unilamellarity will result in a lower encapsulation efficiency than calculated. The theoretical value can be helpful in optimizing the preparation procedure.

[6] Plasmid DNA Vaccines: Entrapment into Liposomes by Dehydration–Rehydration

By GREGORY GREGORIADIS, ANDREW BACON, WILSON CAPARRÓS-WANDERLEY and BRENDA MCCORMACK

I. Introduction

An exciting development in vaccinology is the use of plasmid DNA to elicit humoral and cell-mediated immune responses against the antigen encoded by a gene in the plasmid.[1–3] It is thought[2,3] that immunity follows DNA uptake by muscle cells after intramuscular injection, leading to the expression and extracellular release of the antigen, which is then taken up by antigen-presenting cells (APCs). It is also feasible that some of the injected DNA is taken up directly by APCs infiltrating the site of injection. Disadvantages[1-3] of DNA vaccination include uptake of DNA by only a minor fraction of muscle cells, which at any rate are not professional APCs, exposure of DNA to deoxyribonuclease in the interstitial fluid, thus necessitating the use of relatively large quantities of DNA, and the need in some cases to inject into regenerating muscle in order to enhance immunity. It has been proposed[1,4] that DNA immunization via liposomes (phospholipid vesicles) would protect DNA from nuclease attack[5] and also circumvent muscle cell involvement and facilitate[6] instead the uptake of DNA by APCs infiltrating the site of injection or in the lymphatics, where many liposomes will end up. Moreover, transfection of APCs with liposomal DNA could be promoted by the judicial choice of vesicle characteristics, for instance, surface charge, size, and lipid composition, or by the co-entrapment, together with DNA, of plasmids expressing cytokines (e.g., interleukin 2), immunostimulatory sequences, or appropriate proteins.

To that end, we have developed a method[5,7] (Fig. 1A) that leads to the quantitative entrapment of a number of plasmid DNAs of various sizes into neutral, anionic, and cationic liposomes that are capable of transfecting cells *in vitro* with various levels of efficiency.[5] We have also shown by using this technology that immunization of BALB/c and outbred (T.O.) mice by a variety of routes with (cationic) liposomal DNA leads to much greater humoral [IgG subclasses and splenic interleukin 4 (IL-4)] and cell-mediated [splenic interferon γ (IFN-γ)] immune responses[1,4,7,8] (Fig. 1B, 1D) as well as cytotoxic T lymphocyte responses[9] (Fig. 1D) than those obtained with naked DNA or, in some of the experiments,[4] DNA complexed to preformed similar liposomes. A much greater IgA response was also seen in mice immunized by the oral route with liposomal plasmid DNA[10] (Fig. 1C).

The mechanism by which liposomes promote greater immune responses to the encoded antigen than seen with the naked plasmid is not

[1] G. Gregoriadis, *Pharm. Res.* **15,** 661 (1998).
[2] H. L. Davis, R. G. Whalen, and B. A. Demeneix, B. A. *Hum. Gene Ther.* **4,** 151 (1993).
[3] P. J. Lewis and L. A. Babiuk, *Adv. Virus Res.* **54,** 129 (1999).
[4] G. Gregoriadis, R. Saffie, and J. B. de Souza, *FEBS Lett.* **402,** 107 (1997).
[5] G. Gregoriadis, R. Saffie, and S. L. Hart, *J. Drug Target.* **3,** 469 (1996).
[6] G. Gregoriadis, *Trends Biotechnol.* **13,** 527 (1995).
[7] Y. Perrie and G. Gregoriadis, *Biochim. Biophys. Acta* **1475,** 125 (2000).
[8] Y. Perrie, P. M. Frederik, and G. Gregoriadis, *Vaccine* **19,** 3301 (2001).
[9] A. Bacon, W. Caparros-Wanderley, B. Zadi, and G. Gregoriadis, *J. Liposome Res.* **12,** 73 (2002).
[10] Y. Perrie, M. Obrenovic, D. McCarthy, and G. Gregoriadis, *J. Liposome Res.* **12,** 185 (2002).

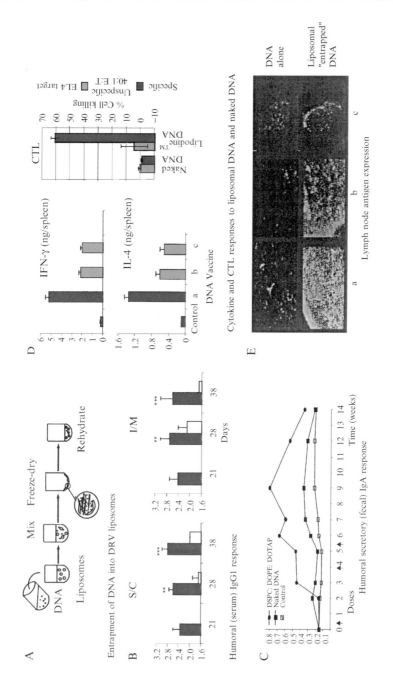

likely to involve muscle cells. Although cationic liposomes could in theory bind to the negatively charged myocytes and be taken up by them, protein in the interstitial fluid would neutralize[6] the liposomal surface and thus interfere with such binding. Moreover, vesicle size (about 600–700 nm, average diameter[7]) would render access to the cells difficult if not impossible. It is then more likely that cationic liposomes are endocytosed by APCs, possibly including dendritic cells, in the lymphatics, where liposomes eventually end up.[6] This is strongly supported by experiments in which mice were injected intramuscularly or subcutaneously with liposomes entrapping plasmid (pCMV.EGFP) encoding enhanced fluorescent green protein or with naked plasmid. Sections of the lymph nodes draining the injected site revealed (Fig. 1E) much more green fluorescence when the plasmid was administered in the entrapped form.[8] Similar observations were made with liposomes given intragastrically (Fig. 1E).[10]

Here we describe methodology for the incorporation of plasmid DNA into liposomes of various lipid composition, vesicle size, and surface charge.

II. Materials

Plasmid DNAs used are pRc/CMV HBS encoding the S (small) protein of the hepatitis B virus surface antigen (HBsAg, subtype ayw),[4] pGL2 encoding luciferase,[5] pRSVGH encoding human growth hormone, pCMV 4.65 encoding *Mycobacterium* leprosy protein, pCMV 4.EGFP encoding enhanced green fluorescent protein, VR 1020 encoding schistosome

---

Fig. 1. (A) Procedure for DNA entrapment into dehydration–rehydration vesicle (DRV) liposomes. (B) Humoral IgG$_1$ response, anti-HBS (hepatitis B surface antigen, S region) titer, after immunization with pRc/CMV HBS. Mice were injected intramuscularly (I/M) or subcutaneously (S/C) on days 1 and 21 with 10 μg of liposome-entrapped (shaded columns) or naked (open columns) DNA and bled on days 21, 28, and 38. p Values: $^{**}p$ <0.005; $^{***}p$ <0.008. (C) Secreted IgA response in mice dosed intragastrically with free or liposome-entrapped pRc/CMV HBS (DNA dose, 100 μg). Response assayed on fecal matter (100 mg/ml) suspended in PBS and clarified by centrifugation. Control mice received only PBS. (D) Cytokine levels in the spleens of mice immunized with DNA entrapped into cationic liposomes (columns a), complexed with cationic (DOTAP) liposomes (columns b), or in naked form (columns c). "Control" denotes cytokine levels in normal nonimmunized mice. Cytokine values in mice immunized with cationic liposomes were significantly higher than those in the other groups (p <0.001–0.05). CTL: CTL response to EL4 cells (targets) loaded with an OVA CTL epitope, or an irrelevant antigen CTL epitope, in mice immunized with pCI-OVA DNA (DNA dose, 10 μg) either alone or entrapped in cationic liposomes (Lipodine). (E) Fluorescence emission images of lymph node sections from mice dosed (a) intramuscularly, (b) subcutaneously (popliteal lymph nodes), or (c) intragastrically (mesenteric nodes) with naked or liposome-entrapped plasmid DNA expressing the enhanced green fluorescent protein (EGFP). Modified from Refs. 1, 4, 8, 9, and 10.

protein, pCI-OVA encoding ovalbumin, and p1.17/SichHA encoding the hemagglutinin antigen of influenza virus (Sichuan strain).

Lipids used include egg phosphatidylcholine (PC), distearoyl phosphatidylcholine (DSPC), egg phosphatidylethanolamine (PE), phosphatidic acid (PA), phosphatidyl glycerol (PG), and phosphatidylserine (PS) (more than 99% pure), all from Lipoid (Ludwigshafen, Germany). Dioleoyl phosphatidylcholine (DOPE) and stearylamine (SA) are from Sigma (Poole, Dorset, UK). The sources of 1,2-bis(hexadecylcycloxy)-3-trimethylaminopropane (BisHOP), N[1-(2,3-dioleoyloxy)propyl]-N,N,N-triethylammonium (DOTMA), 1,2-dioleoyloxy-3-(trimethylammonium propane) (DOTAP), 1,2-dioleoyl-3-dimethylammonium propane (DODAB), and $3\beta(N,N,$-dimethylaminoethane)-carbamyl cholesterol (DC-CHOL) have been described elsewhere.[4,6,8,9] Sepharose CL-4B and polyethylene glycol 6000 are obtained from Pharmacia (Uppsala, Sweden). All other reagents are of analytical grade.

III. Entrapment of Plasmid DNA into Liposomes by Dehydration–Rehydration Procedure

The dehydration–rehydration procedure (Fig. 1A) described below is mild and thus compatible with most labile materials, and easy to perform.

A. Solutions

PC (16 μmol) and DOPE (or PE) (8 μmol) are dissolved in chloroform (2–5 ml). For charged liposomes, 4 μmol of PA, PG, or PS (anionic) or 4 μmol of SA, BisHOP, DOTMA, DOTAP, DODAB, or DC-CHOL (cationic) is also added. The quantity of charged lipids can be increased or decreased depending on the amount of vesicle surface charge required.

Plasmid DNA (up to 500 μg for the amount of PC shown above) is dissolved in 2 ml of distilled water, or 10 mM sodium phosphate buffer, pH 7.2 (phosphate buffer; PB) if needed. The nature of buffer with respect to composition, pH, and molarity can be varied as long as this does not interfere with liposome formation or yield of DNA entrapment. The quantity of added DNA can be increased proportionally to the total amount of lipid used. For cationic liposomes, the amount of added DNA can also be increased by employing additional cationic lipid.

B. Procedure

To entrap plasmid DNA into liposomes (see Fig. 1A) a lipid film is prepared initially from which multilamellar vesicles (MLVs) and, eventually, small unilamellar vesicles (SUVs) are produced. SUVs are then mixed with the plasmid DNA destined for entrapment and dehydrated. Subsequently, the dry cake is broken up and rehydrated to generate multilamellar "dehydration–rehydration" vesicles (DRVs) incorporating the plasmid DNA. On centrifugation, liposome-entrapped DNA is separated from nonentrapped DNA. When required, DNA-containing DRVs are reduced in size by microfluidization in the presence or absence of nonentrapped DNA. Alternatively, small liposomes are produced by employing a novel alternative method[11] that utilizes sucrose (see p. 77).

1. Preparation of Lipid Film

1. The chloroform solution of lipids (Section III.A) is placed in a 50-ml round-bottomed spherical Quick-fit flask. (VWR International LTD., Leichester, U.K.)
2. After evaporation of the solvent in a rotary evaporator at about 37°, a thin lipid film is formed on the walls of the flask.
3. The film is flushed for about 60 s with oxygen-free nitrogen (N_2) to ensure complete solvent removal and to replace air.

2. Preparation of MLVs

1. Distilled water (H_2O, 2 ml) and a few glass beads are added to the flask, the stopper is replaced, and the flask is shaken vigorously by hand or mechanically until the lipid film has been transformed into a milky suspension.
2. This process is carried out above the liquid-crystalline transition temperature (T_c) of the phospholipid ($>T_c$) by prewarming the water before its placement into a prewarmed flask.
3. The suspension is allowed to stand at $>T_c$ for about 1–2 h, whereupon multilamellar liposomes are formed.

3. Preparation of SUVs

1. After the removal of the glass beads, the milky suspension is sonicated at $>T_c$ (with intervals of rest), using a titanium probe slightly immersed into the suspension, which is under N_2 (achieved by the continuous delivery of a gentle stream of N_2 through thin plastic tubing). This step produces a slightly opaque to clear suspension of SUVs (up to about 80 nm in diameter).

[11] B. Zadi and G. Gregoriadis, *J. Liposome Res.* **10,** 73 (2000).

2. The time required to produce SUVs varies, depending on the amount of lipid used and the diameter of the probe. For the amounts of lipid mentioned previously, a clear or slightly opaque suspension is usually obtained within up to four sonication cycles, each lasting 30 s with 30-s rest intervals in between, using a probe 0.75 in. in diameter.

3. The process of sonication is considered successful when adjustment of the settings in the sonicator is such that the suspension is agitated vigorously.

4. The sonicated suspension of SUVs is centrifuged for 2 min at 900 g to remove titanium fragments and the supernatant is allowed to rest at $>T_c$ for about 1–2 h.

4. Dehydration of SUVs in Presence of Added Plasmid DNA

1. SUVs prepared as in Section III.B.3, step 4 are mixed with DNA solution (Section III.A), rapidly frozen in liquid nitrogen while the flask is rotated, and freeze-dried overnight under vacuum (<0.1 torr) in a freeze-dryer.

2. If needed, the suspension can be transferred into an alternative Pyrex container before freezing and drying.

5. Rehydration of Freeze-Dried Material

1. Water (0.1 ml per 16 μmol of PC; prewarmed at $>T_c$) is added to the freeze-dried material and the mixture is swirled vigorously at $>T_c$. The volume of H_2O added must be kept at a minimum, that is, enough H_2O to ensure complete hydration of the powder under vigorous swirling.

2. The suspension is kept at $>T_c$ for about 30 min.

3. The process is repeated with 0.1 ml of H_2O and, 30 min later at $>T_c$, with 0.8 ml of PB (prewarmed at $>T_c$).

4. The suspension is then allowed to stand for about 30 min at $>T_c$.

6. Removal of Nonentrapped DNA

1. The suspension, now containing multilamellar DRVs with entrapped and nonentrapped plasmid DNA, is centrifuged at 40,000 g for 60 min ($4°$).

2. The precipitate (pellet) obtained (DNA-containing DRVs) is suspended in H_2O (or PB) and centrifuged again under the same conditions. The process is repeated at least once to remove the remaining nonentrapped material.

3. The final pellet is suspended in an appropriate volume (e.g., 2 ml) of H_2O or PB. When the liposomal suspension is destined for *in vivo* use (e.g., intramuscular or subcutaneous injection), NaCl is added to a final concentration of 0.9%.

4. The z-average mean diameter of the suspended vesicles measured by photon correlation spectroscopy (PCS) is about 600–700 nm.[7]

IV. Preparation of DNA-Containing Small DRV Liposomes

The following procedures[12] are needed when DNA-containing DRV liposomes are to be converted to smaller vesicles (down to a z-average mean diameter of about 100 nm), still retaining a considerable proportion of the DNA, most of it in the case of cationic DRVs.

A. Microfluidization of DRVs Containing Plasmid DNA

The suspension of liposomes obtained as described in Section III.B.5, step 4 (before separation of entrapped from nonentrapped drug described in Section III.B.6) ("unwashed liposomes") is diluted to 10 ml with H_2O and passed for a number of full cycles through a Microfluidizer 110S (Microfluidics, Newton, MA). The pressure gauge is set at 60 lb/in^2 throughout the procedure to give a flow rate of 35 ml/min.

The number of cycles used depends on the vesicle size required or on the sensitivity of the plasmid DNA. In the case of pGL2, for instance, microfluidization for more than three cycles results in progressive smearing of the DNA and failure to transfect cells *in vitro*.[5] It is likely that other plasmid DNAs will behave similarly on extensive microfluidization.

Microfluidization of the sample can also be carried out after the removal of nonentrapped DNA as described in Section III.B.6 ("washed liposomes"), although DNA retention in this case may be reduced: the presence of unentrapped DNA during microfluidization (a process that destabilizes liposomes, which then reform as smaller vesicles) is expected[12] to diminish DNA leakage, perhaps by reducing the osmotic rupture of vesicles.[12,13] With cationic DRVs, however, DNA is unlikely to leak significantly as it is associated with the cationic charges of the bilayers.

B. Preparation of DNA-Containing Small Liposomes by Sucrose Method

Quantitative entrapment of DNA into small liposomes (up to about 200 nm in diameter) in the absence of microfluidization can be carried out by a novel one-step method[11] as follows.

[12] G. Gregoriadis, H. da Silva, and A. T. Florence, *Int. J. Pharm.* **65**, 235 (1990).
[13] C. Kirby and G. Gregoriadis, *Biotechnology* **2**, 979 (1984).

1. SUVs (e.g., cationic) prepared as described in Section III.B.3 are mixed with sucrose to give a weight:weight ratio of 1.0 (e.g., 21 mg of total lipid for 21 mg of sucrose) and the appropriate amount of plasmid DNA (e.g., 10–500 μg). Alternative lipid-to-sucrose weight ratios can be used to satisfy particular needs (see Zadi and Gregoriadis[11]).

2. The mixture is then frozen and dehydrated by freeze-drying. Rehydration of the cake obtained is carried out as in Section III.B.5.3. For the separation of entrapped from nonentrapped DNA the suspension is centrifuged as in Section III.B.6.

C. Separation of Entrapped from Nonentrapped DNA

1. In the case of microfluidized liposomes, on completion of the number of cycles required, the microfluidized sample (about 10 ml) can, if needed, be reduced in volume by placing the sample in dialysis tubing, which is then covered in a flat container with polyethylene glycol 6000 flakes. As removal of excess H_2O from the tubing is relatively rapid, it is essential that the sample be inspected regularly.

2. When the required volume has been reached, the sample is treated for the separation of entrapped from free DNA. This is carried out either by molecular sieve chromatography, using a Sepharose CL-4B column, or, for cationic liposomes, by centrifugation as in Section III.B.6.

V. Estimation of DNA Entrapment

A. DRVs and Small Sucrose Liposomes

1. Entrapment of plasmid DNA in DRV or small ("sucrose") liposomes is monitored by measuring the DNA in the suspended pellet and combined supernatants. The most convenient way to monitor DNA entrapment is by using radiolabeled (^{32}P or ^{35}S) DNA.

2. If a radiolabel is not available or cannot be used, appropriate quantitative techniques should be employed. To determine DNA by such techniques a sample of the liposomes suspension is mixed with Triton X-100 (up to 5% final concentration) or, preferably, with isopropanol (1:1 volume ratio) so as to liberate the DNA.

3. However, if Triton X-100 or the solubilized liposomal lipids interfere with the assay of the DNA, liposomal lipids or the DNA must be extracted by appropriate techniques (e.g., Gregoriadis et al.[5]). Entrapment values range between about 30 and 99%, depending on the DNA used and the presence or absence of a cationic charge. Values are highest when DNA is entrapped into cationic DRVs (Table I).

TABLE I
INCORPORATION OF PLASMID DNA INTO LIPOSOMES BY DEHYDRATION–REHYDRATION METHOD

Liposomes	Mode of incorporation[c]	Incorporated plasmid DNA (% of plasmid used)[a,b]							
		pGL2	pRc/CMV/HBS	pRSVGH	pCMV 4.65	pCMV 4.EGFP	VR 1020	pCI-OVA	p1.17/SichHA
PC, DOPE	*	44.2[d]	55.4 (19.3)[e]	45.6	28.6				
	†	12.1		11.3					
PC, DOPE, PS	*	57.3[d]							
	†	12.6							
PC, DOPE, PG	*			53.5					
	†			10.2					
PC, DOPE, SA	*	74.8[d]							
	†	48.3							
PC, DOPE, BisHOP	*	69.3[d]							
PC, DOPE, DOTMA	*	86.8							
PC, DOPE, DC-CHOL	*		87.1	76.9					
	†			77.2					
PC, DOPE, DOTAP	*		80.1 (96.7)[e]	79.8	52.7	71.9	89.6	91.4	98.6 (93.0)[e]
	†		88.6	80.6	67.7		81.6		
PC, DOPE, DODAP	*			57.4					
	†			64.8					

[a] Plasmid DNAs used encoded luciferase (pGL2), hepatitis B surface antigen (S region) (pRc/CMV HBS), human growth hormone (pRSVGH), *Mycobacterium leprosy* protein (pCMV 4.65), green fluorescent protein (pCMV 4.EGFP), schistosome protein (VR 1020), ovalbumin (pCI-OVA), and hemagglutinin antigen (p1.17/SichHA).

[b] Incorporation values for the various amounts of DNA used for each of the liposomal formulations did not differ significantly and were therefore pooled (values shown are means of values obtained from three to five experiments). PC (16 μmol) was used in molar ratios of 1:0.5 (neutral) and 1:0.5:0.25 (anionic and cationic liposomes).

[c] ^{35}S-Labeled plasmid DNA (10–500 μg) was incorporated (*) into or mixed (†) with neutral (PC, DOPE), anionic (PC, DOPE, PS, or PG), or cationic (PC, DOPE, SA, BisHOP, DOTMA, DC-CHOL, DOTAP, or DODAP) dehydration–rehydration vesicles (DRVs).

[d] Entrapment values for microfluidized DRVs were 12–83%, depending on the vesicle charge and amount of DNA used.[5]

[e] Entrapment values in parentheses were obtained by the dehydration–rehydration method carried out in the presence of sucrose (see text). z-Average vesicle size was 200 nm (SichHA), 180 nm (HBS, cationic), and 90 nm (HBS, neutral).

B. Measurement of DNA Retention by Microfluidized DRVs and Vesicle Characteristics

1. The content of DNA within liposomes is estimated as in Section V.A and expressed as a percentage of the DNA in the original preparation obtained as described in Section III.B.6.

2. When the sample is microfluidized according to the protocol in Section IV.A, that is, before the estimation of entrapment, it is necessary that a small portion of the sample to be microfluidized be kept aside for the estimation of entrapment according to the protocol in Section V.

3. Vesicle size measurements are carried out by photon correlation spectroscopy as described elsewhere.[5,7,12,14]

4. Liposomes with entrapped DNA can also be subjected to microelectrophoresis in a Zetasizer (Malvern Instruments, Southboro, MA) to determine their ζ potential. This is often required in order to determine the net surface charge of DNA-containing cationic liposomes.[7]

[14] N. Skalko, J. Bouwstra, F. Spies, and G. Gregoriadis, *Biochim. Biophys. Acta* **1301**, 249 (1996).

[7] Interdigitation–Fusion Liposomes

By Patrick L. Ahl and Walter R. Perkins

Interdigitation–Fusion Process

A closed vesicle formed by one or more lipid bilayers is perhaps the simplest description of a liposome. This description encompasses the basic idea that liposomes have an interior compartment that is separated from the external medium by a permeability barrier. The ability to encapsulate molecules either in the interior compartment or within the lipid bilayer is the starting point for using liposomes as drug delivery vehicles. Liposome-based drug delivery technology has advanced well beyond the laboratory and several liposome-based therapeutics are now on the market.[1–3] Liposomes are also valuable research tools. They are essential for examining the structure and function of isolated membrane proteins, the biophysical aspects of biomembrane fusion, and the biomembrane-related events of signal transduction. The composition and physical structure of liposomes are as diverse as the applications discussed above. This chapter describes how to prepare novel liposomes with extremely high internal volumes. These liposomes are formed by interdigitation-driven fusion of small unilamellar vesicles (SUVs). Conditions that promote bi-layer interdigitation such as ethanol[4] or hydrostatic pressure[5] first cause the SUVs to fuse into large sheets of interdigitated lamellar phospholipids, which then transform into exceptionally large and unilamellar liposomes when the temperature is raised above the liquid-crystalline phase transition temperature (T_m) of the phospholipid. We refer to the entire process as the interdigitation–fusion (IF) method and the product liposomes as inter-digitation–fusion liposomes, or IF-liposomes. IF-liposomes may also be referred to as interdigitation–fusion vesicles (IFVs), as in our original pub-lication.[4] The phospholipid membranes of the IF-liposomes are not inter-digitated, because after the IF-liposomes are formed the interdigitating conditions are removed and the liposome phospholipid returns to the normal bilayer state. The interdigitation–fusion procedure is a useful and patented method for preparing large quantities of essentially unilamellar liposomes with internal volumes that can exceed 25 $\mu l/\mu mol$ phospholipid.[6]

The principal limitation of the IF process is that IF-liposome formula-tions must use phospholipids that can form an interdigitated bilayer, that is, the $L_{\beta I}$ phase. Full interdigitation requires that the hydrocarbons of the fatty acid chains be essentially straight in the all-*trans* conformation. Many commonly used phospholipids do not form an $L_{\beta I}$ phase. Phospholipids with acyl chains containing permanent "kinks" due to *cis* double bonds such as 1,2-dioleoyl-*sn*-glycero-3-phosphocholine (DOPC), 1-palmitoyl-2-oleoyl-*sn*-glycero-3-phosphocholine (POPC), or egg phosphatidylcholine (EPC) do not interdigitate and will not form IF-liposomes. Likewise, phospholipids at temperatures above their T_m in the liquid-crystalline phase, that is, the L_α phase, will also not form IF-liposomes. Bilayer inter-digitation is also unfavorable for phospholipids with small polar head groups relative to the hydrophobic portion of the molecule, such as phosphatidylethanolamines. The interdigitation–fusion method functions

[1] P. G. Schmidt, J. P. Adler-Moore, E. A. Forssen, and R. T. Proffitt, *in* "Medical Applications of Liposomes" (D. D. Lasic and D. Papahadjopoulos, eds.), p. 703. Elsevier Science, Amsterdam, 1998.

[2] F. J. Martin, *in* "Medical Applications of Liposomes" (D. D. Lasic and D. Papahadjopou-los, eds.), p. 635. Elsevier Science, Amsterdam, 1998.

[3] C. E. Swenson, J. Freitag, and A. S. Janoff, *in* "Medical Applications of Liposomes" (D. D. Lasic and D. Papahadjopoulos, eds.), p. 689. Elsevier Science, Amsterdam, 1998.

[4] P. L. Ahl, L. Chen, W. R. Perkins, S. R. Minchey, L. T. Boni, T. F. Taraschi, and A. S. Janoff, *Biochim. Biophys. Acta* **1195**, 237 (1994).

[5] W. R. Perkins, R. Dause, L. Xingong, T. S. Davis, P. L. Ahl, S. R. Minchey, T. F. Taraschi, S. Erramilli, S. M. Gruner, and A. S. Janoff, *J. Liposome Res.* **5**, 605 (1995).

[6] L. T. Boni, A. S. Janoff, S. R. Minchey, W. R. Perkins, C. E. Swenson, and T. S. Davis, unpublished data (1993).

best for phosphatidylcholines with fully saturated acyl chain, such as 1,2-dipalmitoyl-*sn*-glycero-3-phosphocholine (DPPC) or 1,2-distearoyl-*sn*-glycero-3-phosphocholine (DSPC), that are below the T_m of the phospholipid. DPPC or DSPC IF-liposome formulations have high internal volumes typically exceeding 20 $\mu l/\mu$mol lipid. Thus, DPPC or DSPC IF-liposome formulations are ideal for medical applications that require liposomes with a high internal volume.

Ethanol-induced interdigitation occurs when ethanol replaces water molecules at the hydrocarbon–water interface of gel-phase ($L_{\beta'}$ phase) phosphatidylcholine bilayers. Ethanol is a larger molecule than water and increases the lateral separation between phospholipid head groups. This increased lateral separation between the phospholipid head groups would produce energetically unfavorable voids within the hydrocarbon region of the noninterdigitated bilayer. This problem can be avoided by sliding the hydrocarbon chains from opposing bilayers past each other, creating the interdigitated bilayer, that is, the $L_{\beta I}$ phase.[7] Simple geometry implies that such a rearrangement of the phospholipids would reduce the thickness of the membrane. Ethanol-induced reductions in membrane thickness consistent with phospholipid interdigitation were confirmed by small-angle x-ray diffraction measurements on DPPC multilamellar vesicles (MLVs). The MLVs are prepared by simple hydration of dried lipid films.[8] Small-angle x-ray diffraction from MLVs that have a large amount of stacked planar membranes allows for calculation of the thickness and electron density profile of the membrane. This is probably the most reliable test for bilayer interdigitation.

Interdigitated phospholipid membranes transform into the noninterdigitated L_α phase when the temperature is raised above the T_m of the phospholipid. The L_α phase can accommodate the increased lipid head group separation produced by the ethanol because of the fluid nature of the hydrocarbon chains. Nambi *et al.*[9] used the fluorescent probe 1,6-diphenyl-1,3,5-hexatriene (DPH) to establish the temperature–ethanol phase diagrams for DPPC and DSPC MLVs. The room temperature $L_{\beta'}$-to-$L_{\beta I}$ phase transition occurs around 1.0 M ethanol for MLVs of these saturated phosphatidylcholines. Essentially all the initial research with interdigitated phospholipid membranes was done with MLVs, because MLVs are easy to prepare and analyze by many biophysical techniques such as differential scanning calorimetry and X-ray diffraction. The planar bilayer structure

[7] S. A. Simon and T. J. McIntosh, *Biochim. Biophys. Acta* **773**, 169 (1984).
[8] W. R. Perkins, S. R. Minchey, M. J. Ostro, T. F. Taraschi, and A. S. Janoff, *Biochim. Biophys. Acta* **943**, 103 (1988).
[9] P. Nambi, E. S. Rowe, and T. J. McIntosh, *Biochemistry* **27**, 9175 (1988).

of MLV membranes also favors the formation of the interdigitated membrane. It is easy to imagine how the hydrocarbon tails of planar opposing lipid monolayers could slide by each other with little steric interference. However, if there is extreme bending of the opposing phospholipid monolayers, then the individual lipids might not be able to so easily slide pass each other and bilayer interdigitation would be inhibited.

The question of the relationship between interdigitation and membrane curvature led Boni et al.[10] to measure ethanol-induced interdigitation of DPPC liposomes as a function of liposome diameter. Consistent with the previous discussion, approximately 0.8 M ethanol was sufficient to induce room temperature interdigitation of DPPC liposomes extruded through 1000-nm pore size polycarbonate filters, as measured by the fluorescent probe DPH. In contrast, over 1.2 M ethanol was required for DPPC liposomes extruded through 100-nm pore size filters. Because small-angle X-ray measurements are typically not capable of determining the membrane electron density profile of unilamellar liposomes, wide-angle X-ray scattering (WAXS) was used to identify the onset of the interdigitated phase for SUVs. Wide-angle X-ray scattering arises from distance correlations between the hydrocarbon chains within the bilayer and thus is insensitive to lamellarity.[11] The WAXS peak (4.2 Å) for $L_{\beta'}$ DPPC MLVs shifts lower (4.09 Å) when the interdigitated state is induced by high pressure because interdigitation results in a tighter packing of the hydrocarbon chains. Figure 1 shows how this WAXS peak was used to determine the onset of ethanol-induced interdigitation for DPPC MLVs versus DPPC SUVs at room temperature. The WAXS peak shift corresponding to the $L_{\beta'}$-to-$L_{\beta I}$ transition occurred between 0.75 and 1.0 M ethanol for the DPPC MLV sample. In contrast, the broad WAXS peak corresponding to the noninterdigitated state of the DPPC SUVs (4.13 Å) did not shift to the position of the $L_{\beta I}$ WAXS peak until the ethanol concentration was over 1.5 M. Thus, interdigitation was clearly inhibited in the highly curved SUV bilayer and required more ethanol. The free energy of SUV interdigitation is so unfavorable that higher amounts of ethanol are required. When interdigitation does proceed, it produces deformed SUVs that have membrane defects where hydrophobic domains are exposed.[10] The SUVs then fuse together at these hydrophobic defects into large extended planar sheets of interdigitated phospholipid. Ultimately, the IF process proceeds because the $L_{\beta I}$ lipid sheet phase has a lower

[10] L. T. Boni, S. R. Minchey, W. R. Perkins, P. L. Ahl, J. L. Slater, M. W. Tate, S. M. Gruner, and A. S. Janoff, Biochim. Biophys. Acta **1146**, 247 (1993).
[11] L. F. Braganza and D. L. Worcester, Biochemistry **25**, 2591 (1986).

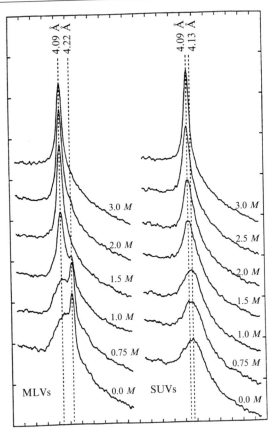

FIG. 1. Wide-angle x-ray scattering (WAXS) intensities of DPPC MLVs (*left*) and SUVs (*right*) after the addition of ethanol at 25°. The ethanol concentrations are indicated to the right of each curve. The shift in the 4.22-Å MLV diffraction peak to 4.09 Å occurs between 0.75 and 1.0 M ethanol and is due to the $L_{\beta'}$-to-$L_{\beta I}$ phase transition. The DPPC SUVs have a broad diffraction peak at 4.13 Å that begins to sharpen and shift to 4.09 Å above 1.5 M ethanol. The MLV and SUV WAXS peaks at 3.0 M ethanol are equivalent. From L. T. Boni *et al.*, *Biochim. Biophys. Acta* **1146**, 247 (1993).

free energy than the SUVs under these conditions of ethanol concentration and temperature.

Concentrations of ethanol sufficient to induce interdigitation almost immediately transform the relatively clear suspensions of DPPC or DSPC SUVs into opaque and extremely viscous suspensions of interdigitated sheets that can extend tens of micrometers in length. A phase-contrast microscopy photograph of DPPC interdigitated sheets formed from DPPC SUVs in

3.0 M ethanol is shown in Fig. 2A. Figure 2B shows a corresponding freeze–fracture electron microscopy (EM) photograph of DPPC interdigitated sheets in 3.0 M ethanol. As discussed above, the interdigitated $L_{\beta I}$ phase requires that the fatty acid hydrocarbon chains be in the all-*trans* conformation. Raising the temperature above the phospholipid T_m converts the lipid sheets from the $L_{\beta I}$ phase to the liquid-crystalline, that is, L_α phase. The L_α phase does not restrict membrane curvature. The lipid sheets can bend back on themselves above the T_m to form the closed vesicles we refer to as IF-liposomes. The process of vesicularization occurs because it eliminates the energetically unfavorable lipid–water interfaces at the edges of the lipid sheets. The vesicularization process results in the production of large and highly unilamellar liposomes, probably because of the extended nature of the lipid sheets. Once the vesicles are formed, lowering the temperature below the T_m of the lipid does not reform the phospholipid sheets because the IF-liposome membranes are no longer in the highly curved condition. The IF-liposome bilayers will still reinterdigitate, if sufficient ethanol is present, but the lipid sheets will not reform. Unlike in the case of SUVs, however, liposome fusion is not required to sustain the interdigitation of large liposomes. Removing the ethanol after vesicle formation leaves large unilamellar IF-liposomes with high internal volumes and normal phospholipid bilayers.[4,10]

Braganza and Worcester found that hydrostatic pressure can also induce lipid bilayer interdigitation.[11] The $L_{\beta I}$ phase must have a lower free energy than $L_{\beta'}$ under high hydrostatic pressure because of denser hydrocarbon packing in $L_{\beta I}$. Perkins et al.[5] used hydrostatic pressure to prepare IF-liposomes without the use of the interdigitation inducer ethanol. DPPC or DSPC SUVs were put under hydrostatic pressure sufficient to induce lipid interdigitation (>2 kbar). A significant proportion of the SUVs fused under those conditions in order to convert to the lower free energy state, which is the $L_{\beta I}$ phase. The SUVs vesicularized into large unilamellar liposomes if the sample was maintained at sufficiently high temperature when the hydrostatic pressure was removed. We refer to these IF-liposomes as pressure-induced fusion liposomes or PIF-liposomes in order to differentiate them from ethanol-induced IF-liposomes. Repeated cycles of hydrostatic pressure were found to increase significantly the internal volume of the PIF-liposomes. Internal volumes of 27–37 $\mu l/\mu mol$ phospholipid were attained for DSPC PIF-liposomes after six cycles to 4 kbar at 54°. In addition to the benefit of avoiding ethanol, the proper application of the PIF process produced sterile samples.[5] A practical disadvantage of the PIF process is that it requires a hydrostatic pressure apparatus capable of producing pressures in the 3- to 6-kbar range. Unless such equipment is readily available, ethanol-induced SUV interdigitation fusion is generally a much more convenient procedure for preparing IF-liposomes.

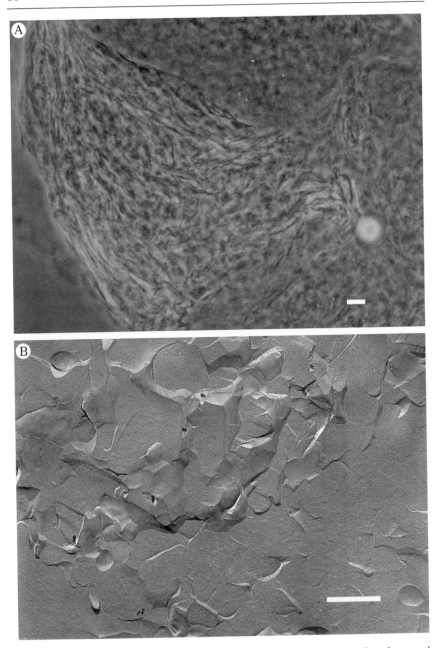

Fig. 2. Appearance of DPPC interdigated sheets. (A) Phase-contrast microphotograph of DPPC interdigitated sheets after the room temperature addition of 3.0 M ethanol. The DPPC interdigitated sheets were at 20 mg/ml in NaCl–Tris buffer. Scale bar: 10 μm.

Ethanol-induced interdigitation–fusion process

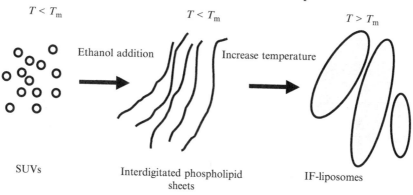

FIG. 3. Schematic outline of ethanol-induced IF-liposome formation through the interdigitation–fusion of SUVs. See text for details.

Preparation of Ethanol-Induced Interdigitation–Fusion Liposomes

Preparation of IF-liposomes by ethanol-induced interdigitation–fusion is easy and relatively straightforward. The preparation conditions can be adjusted to meet specific IF-liposome requirements. However, the following general guidelines must be followed: (1) The phospholipids used in the formulation must have an $L_{\beta I}$ phase; (2) the diameter of the precursor liposomes should be as small as possible, preferably sonicated SUVs; (3) the temperature of the SUV suspension after the addition of the ethanol must be below the T_m of the phospholipid; and (4) the temperature should be raised above the T_m of the lipids after the formation of the interdigitated lipid sheets. These important guidelines are schematically illustrated in Fig. 3.

The procedure described here outlines the basic protocol and optimal conditions for producing laboratory-scale preparations of either DPPC or DSPC ethanol-induced IF-liposomes. Ethanol-induced IF-liposomes are typically prepared in 4-ml batches with 4.0 M ethanol. MLV liposomes are first prepared by directly hydrating lyophilized DPPC or DSPC at a typical lipid concentration of 20 mg/ml in 150 mM NaCl, 10 mM Tris–HCl, pH 7.4 (NaCl–Tris buffer). The MLVs are transformed into SUVs by probe sonication above the T_m of the lipid (e.g., DPPC, 50–55°; DSPC,

(B) Freeze–fracture EM photograph of DPPC interdigitated sheets at 20 mg/ml lipid after the room temperature addition of 3.0 M ethanol. Scale bar: 2 μm. The experimental methods can be found in P. L. Ahl *et al.*, *Biochim. Biophys. Acta* **1195,** 237 (1994).

Fig. 4. Appearance of DPPC SUV sample (left-hand vial), interdigitated sheet sample (center vial), and IF-liposome sample (right-hand vial). The DPPC concentration was 20 mg/ml for each case in NaCl–Tris buffer. The DPPC interdigitated sheets were formed by the addition of 3.0 M ethanol at room temperature. The IF-liposomes were formed by raising the temperature to 55° for 15 min. The ethanol was not removed. The DPPC interdigitated sheet sample vial is inverted to demonstrate the extreme viscosity of the sample. (See Color Insert.)

65–70°) until the samples are relatively clear. Titanium dust produced by the sonicator tip is removed by centrifugation. The SUV suspension is transferred to a vial that can be sealed and allowed to cool below the T_m of the phospholipid. Room temperature is suitable for either DSPC or DPPC. A volume of absolute ethanol is added to bring the final ethanol concentration in the sample to 4.0 M. The sample is then sealed immediately, vortexed, and allowed to incubate at room temperature for at least 10 min. This procedure produces a quick and dramatic change in the appearance of the sample. The previously relatively transparent SUV suspension changes to an extremely viscous and opaque white suspension of interdigitated phospholipid sheets. The sample is then transferred to a water bath and incubated for 15 min at a temperature above the T_m of the lipid (e.g., DPPC, 50–55°; DSPC, 65–70°). This incubation step transforms the interdigitated phospholipid sheets to sealed IF-liposomes. The viscosity of the suspension is decreased significantly on vesicularization. These sample transformations are shown in Fig. 4. The left-hand vial shows the appearance of a DPPC SUV sample at 20 mg/ml lipid before the addition of ethanol. The center vial shows the appearance of the interdigitated DPPC sheet sample immediately after the addition of ethanol. The vial has been inverted to demonstrate the extremely viscous nature of the DPPC interdigitated sheet suspension. The right-hand vial shows the appearance of the sample after the high-temperature incubation. At this point the DPPC sheets have converted to vesicles, that is, IF-liposomes. Although the IF-liposome vial is still white and opaque, the sample flows and is no longer extremely viscous, particularly above the lipid T_m. The ethanol is typically removed from the IF-liposome sample at this point by gently bubbling a stream of N_2 gas through the sample at a temperature above the phospholipid T_m. This procedure can be followed by several centrifugation washes at room temperature with the NaCl–Tris buffer (12,000g, 15 min). This procedure results typically in recovery of 90 to 100% of the initial phospholipid because of the large size and highly aggregated state of DPPC and DSPC IF-liposomes. Other IF-liposomes that are smaller in diameter and/or less aggregated than DPPC or DSPC IF-liposomes might be more difficult to pellet using the centrifugation conditions described above. These IF-liposome formulations might require other techniques for complete ethanol removal. For pharmaceutical-scale preparation, sample homogenization works well to produce SUV liposomes, and diafiltration would be a suitable method for ethanol removal. Although the process would require larger scale equipment, the basic procedures would be similar to laboratory-scale preparation.

The principal parameter used to characterize IF-liposomes is internal volume, which is simply the enclosed aqueous volume inside the vesicle divided by the moles of phospholipid. The internal volume of a liposome is also often referred to as the captured volume. Perkins et al.[12] discuss the significance of internal volume and several methods for determining this important liposome parameter in detail. The internal volumes of DPPC and DSPC IF-liposomes have been determined by several methods, but our preferred method is the electron paramagnetic resonance (EPR) ViVo method.[8] This method involves the addition of a precise amount of the membrane-impermeant EPR spin probe 4-trimethylammonium-2,2,6,6-tretramethyl-1-piperidinyloxy (CAT-1) free radical followed by centrifugation to pellet the IF-liposomes. The concentration of the spin probe in the supernatant over that which would be expected in the absence of the IF-liposomes is then determined, using a standard curve. This allows accurate measurement of the external volume of a suspension of large liposomes such as IF-liposomes, which in turn allows determination of the total

[12] W. R. Perkins, S. R. Minchey, P. L. Ahl, and A. S. Janoff, *Chem. Phys. Lipids* **64**, 197 (1993).

TABLE I

INTERNAL VOLUMES OF SATURATED PHOSPHATIDYLCHOLINE INTERDIGITATION–FUSION LIPOSOMES[a]

| | Phospholipid | | | |
	DMPC	DPPC	DSPC	DAPC
Fatty acids (chain length: cis bonds)	di-14:0	di-16:0	di-18:0	di-20:0
Internal volume ($\mu l/\mu mol$)	6.63 ± 0.06	20.23 ± 0.21	23.67 ± 0.06	11.80 ± 0.96

[a] IF-liposomes were prepared, using 4.0 M ethanol, from precursor SUVs at 20 mg/ml. DAPC, 1,2-Diarachidoyl-sn-glycero-3-phosphocholine.
From P. L. Ahl et al., Biochim. Biophys. Acta **1195**, 237 (1994).

volume inside the liposomes. This total volume inside the liposomes is normalized to the amount of phospholipid, using a phosphate assay.[13] Table I shows the internal volumes of IF-liposomes prepared with four saturated phosphatidylcholines, using 4.0 M ethanol. The DMPC (1,2-dimyristoyl-sn-glycero-3-phosphocholine) interdigitated sheets were formed by adding the 4.0 M ethanol at 5° rather than at room temperature in order to be below the T_m of DMPC. Control liposomes formed from phospholipids that do not undergo ethanol-induced interdigitation, such as DOPC or EPC, had internal volumes of less than 1 $\mu l/\mu mol$ lipid when prepared by the same IF procedure. The three main variables in ethanol-induced IF-liposome preparation are ethanol concentration, precursor liposome diameter, and precursor liposome concentration. The effects of each of these variables on the final internal volume of DPPC IF-liposomes are shown in Fig. 5. The optimum conditions for the formation of high internal volume DPPC IF-liposomes are (1) ethanol concentrations near 4.0 M, (2) SUV precursor liposomes, and (3) lipid concentrations around 20 mg/ml. The decrease in internal volume of DPPC IF-liposomes formed at high lipid concentrations was due primarily to an increase in liposome lamellarity rather than to a reduction in liposome size. Statistical lamellarities for DPPC IF-liposome formed at 20 and 100 mg/ml were 1.1 and 2.9 bilayers per vesicle, respectively, as determined by ^{31}P nuclear magnetic resonance (^{31}P NMR).[4,14] The close proximity of the interdigitated phospholipid sheets found at high lipid concentration probably results in IF-liposomes being trapped inside IF-liposomes. This scenario would explain the increase in liposome lamellarity and the reduced internal volume of the IF-liposomes.

[13] G. R. Bartlett, J. Biol. Chem. **234**, 466 (1959).
[14] M. J. Hope, M. B. Bally, G. Webb, and P. R. Cullis, Biochim. Biophys. Acta **812**, 55 (1985).

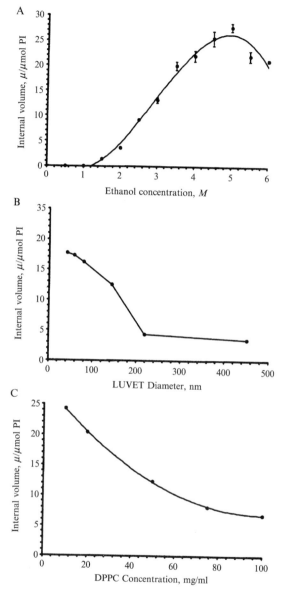

Fig. 5. Effect of ethanol concentration, precursor liposome diameter, and lipid concentration on the internal volume of IF-liposomes. (A) Internal volume of ethanol-induced DPPC IF-liposomes formed from SUVs as a function of ethanol concentration. The ethanol was directly added to samples prepared at 2.0 M ethanol and above. The ethanol was prediluted for samples prepared with ethanol concentrations of 1.5 M and below. The DPPC SUVs were

Preparation of Pressure-Induced Interdigitation–Fusion Liposomes

The preparation of pressure-induced IF-liposomes, that is, PIF-liposomes, requires a high-pressure pump capable of reaching hydrostatic pressures of at least 4 kbar.[5] A suitable pump, reactor cell, and other necessary apparatus can be obtained from High Pressure Equipment Company (Erie, PA). DPPC or DSPC PIF-liposomes are prepared by loading SUVs of either of these lipids into a Teflon tube sample holder fitted at each end with removable caps. The loaded sample holder is then submerged in the hydraulic fluid of the reactor cell and the reactor cell is capped. The temperature of the reactor cell can be brought to the desired temperature by wrapping it with tubing through which water from a temperature-controlled bath is circulated. The reactor cell is then brought to the desired hydrostatic pressure. The pressure is maintained for 15 min, after which the pressure is quickly reduced to atmospheric pressure. High-pressure incubations longer than 15 min did not result in any further improvement in PIF-liposome internal volume.[5] The sample is removed from the reaction cell, transferred to a glass vial, and heated at a temperature approximately 10° above the T_m of the lipid for at least 5 min. The internal volumes (or captured volumes) of DPPC and DSPC PIF-liposomes prepared according to this procedure as a function of temperature and hydrostatic pressure are shown in Fig. 6. A single PIF cycle does not force the fusion of all available SUVs. Freeze–fracture EM photographs of PIF-liposome samples after one cycle revealed precursor SUVs together with large PIF-liposomes.[5] Repeating the pressurization cycle reduced significantly the number of precursor SUVs. This procedure increased the internal volume of the final IF-liposome preparation. DSPC PIF-liposomes with internal volumes in the 27- to 37-$\mu l/\mu mol$ range have been attained after six cycles to 4 kbar at 54°. The PIF process of IF-liposome formation has the added advantage of sample sterilization at relatively low temperatures. Temperatures and hydrostatic pressures ideal for PIF-liposome formation (e.g., 40–60° and 3.4–6.2 kbar) will completely eradicate *Bacillus subtilis* and *Bacillus stearothermophilus* spores.[5]

---

at 20 mg/ml in NaCl–Tris buffer. (B) Internal volume of ethanol-induced DPPC IF-liposomes as a function of the precursor liposome (LUVET) average diameter. The DPPC LUVETs were prepared by 10× extrusion of DPPC MLVs through two polycarbonate filters of various pore sizes. The diameters of the DPPC LUVETs were determined by quasi-elastic light scattering. The DPPC IF-liposomes were formed at 20 mg/ml, using 4.0 M ethanol. (C) The effect of precursor DPPC SUV concentration on the internal volume of ethanol-induced DPPC IF-liposomes. The IF-liposomes were formed at 4.0 M ethanol. From P. L. Ahl *et al.*, *Biochim. Biophys. Acta* **1195**, 237 (1994).

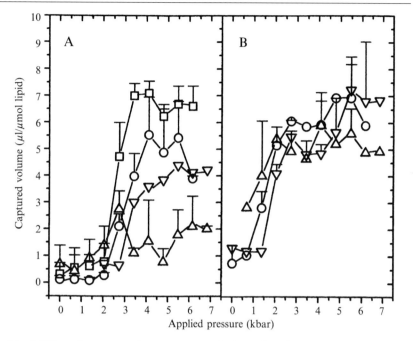

FIG. 6. Effect of pressure and temperature on the formation of PIF-liposomes from DPPC (A) or DSPC (B). The samples were pressurized at the following temperatures: 23° (△); 40° (□); 50° (○); and 64° (▽). Captured volume is another name for liposome internal volume. Error bars represent the standard deviation for $n \geq 3$. Only plus error bars are shown to reduce confusion. SUVs were prepared by sonication at 20 mg/ml. Samples were exposed to the indicated pressure for 15 min at the indicated temperature. The samples were heated to 10° above the lipid T_m after the pressurization step to ensure complete IF-liposome formation. From W. R. Perkins et al., J. Liposome Res. **5,** 605 (1995).

Interdigitation–Fusion Liposome Properties and Applications

IF-liposomes have two properties, large diameters and unilamellar bilayers, that distinguish them from most other liposome types. These features give rise to the exceptionally large internal volume of IF-liposomes. In addition, preparation of highly concentrated suspensions of IF-liposomes is extremely easy, using just ethanol. These characteristics make the ethanol-induced IF process an ideal method for efficiently encapsulating many water-soluble agents. A simple calculation based on the results shown in Fig. 5c demonstrates clearly how efficiently IF-liposomes encapsulate water-soluble material. Because DPPC IF-liposomes have an internal volume of 15 μl/μmol lipid at 40 mg/ml, approximately 75% of the total sample volume is inside the IF-liposomes. Thus, the encapsulation efficiency of any water-soluble material under these conditions should be around 75%. Table II compares the internal volumes of IF-liposomes with the internal volumes of liposomes prepared by a variety of common methods. The internal volume for any given liposome preparation will depend on many factors including lipid composition, lipid concentration, preparation temperature, and other details of any particular process. However, Table II indicates the general range of liposome diameters and internal volumes expected when preparing liposomes by these commonly used methods. Table II illustrates that the internal volumes of IF-liposomes usually exceed significantly the internal volumes of most other liposome types. Internal volumes higher than those of IF-liposomes have been reported for lipid-based structures prepared by the solvent-spherule evaporation method.[15] These structures, which are referred to as DepoFoam vesicles, are large multicompartmental liposomes with diameters in the 5- to 30-μm range and are not suitable for intravenous injection.

In addition to ethanol, other chemical agents have been reported to induce phospholipid interdigitation such as glycerol,[16] methanol,[16] ethylene glycol,[16] benzyl alcohol,[17] chlorpromazine,[17] tetracaine,[17] thiocyanate ion,[18] and polymyxin B.[19,20] The ether-linked phosphatidylcholine DHPC (1,2-O-dihexadecyl-sn-glycero-3-phosphocholine) and some fully saturated mixed hydrocarbon chain length phosphatidylcholines have been observed to interdigitate or partially interdigitate in the absence of any inducer.[21–23] In principle, any of the above-mentioned inducers could be used to prepare IF-liposomes, but so far only ethanol and hydrostatic pressure have been successfully used to make high internal volume IF-liposomes.

The most significant disadvantage to using either the ethanol-induced or the pressure-induced IF process to formulate liposomes is the requirement that the lipid composition must form an interdigitated membrane. As previously discussed, this sometimes limits the choice of the primary phospholipid in any IF-liposome formulation to a phospholipid with all-*trans* hydrocarbon chains. A large majority of the phospholipids in a multicomponent liposome formulation should be capable of bilayer interdigitation for the IF process to be effective. Liposome formulations containing a high

[15] S. Kim, R. E. Jacobs, and S. H. White, *Biochim. Biophys. Acta* **812**, 793 (1985).
[16] R. V. McDaniel, T. J. McIntosh, and S. A. Simon, *Biochim. Biophys. Acta* **731**, 97 (1983).
[17] T. J. McIntosh, R. V. McDaniel, and S. A. Simon, *Biochim. Biophys. Acta* **731**, 109 (1983).
[18] B. A. Cunningham and L. J. Lis, *Biochim. Biophys. Acta* **861**, 237 (1986).
[19] J. M. Boggs and G. Rangaraj, *Biochim. Biophys. Acta* **816**, 221 (1985).
[20] J. L. Ranck and J. F. Tocanne, *FEBS Lett.* **143**, 175 (1982).
[21] M. J. Ruocco, D. J. Siminovitch, and R. G. Griffin, *Biochemistry* **24**, 2406 (1985).
[22] C. Huang and T. J. McIntosh, *Biophys. J.* **72**, 2702 (1997).
[23] S. W. Hui, J. T. Mason, and C. Huang, *Biochemistry* **23**, 5570 (1984).

TABLE II

INTERNAL VOLUMES OF LIPOSOMES PREPARED BY STANDARD PROCEDURES

Liposome type	Preparation description	Diameter (μm)	Internal volume (μl/μmol lipid)	Ref.[a]
IF-liposomes	DSPC or DPPC, 4 M ethanol	1–2	20–25	1
	DPPC–cholesterol (10–35% cholesterol), 4 M ethanol	1–2	8–17	1
PIF-liposomes	DPPG, 4 M ethanol	0.5–2	16	2
	DPPC, one pressure cycle, 4 kbar, 51°	0.5–2	5–7	3
	DSPC, six pressure cycles, 4 kbar, 54°	0.5–2	27–37	3
Detergent dialysis liposomes	Removal of detergent by dialysis	0.2–0.5	3–8	4
Frozen and thawed vesicles (FATMLVs)	Frozen and thawed MLVs	1–5	5–7	5
Reversed phase evaporation vesicles (REVs)	Organic solvent removed by evaporation	0.1–1	8–16	6
Stable plurilamellar lipid vesicles (SPLVs)	Organic solvent removed by stream of N_2 gas	0.1–3	3–4	7
Large unilamellar vesicles by extrusion (LUVETs)	FATMLVs extruded through polycarbonate filter	0.1–0.3	0.9–1.2	8
Multilamellar vesicles (MLVs)	Hydrated from dried lipid film	1–10	0.5–1.2	9
Small unilamellar vesicles (SUVs)	Extensive sonication	0.02–0.03	0.3	10

[a] References: (1) P. L. Ahl et al., Biochim. Biophys. Acta 1195, 237 (1994); (2) W. R. Perkins et al., Chem. Phys. Lipids 64, 197 (1993); (3) W. R. Perkins et al., J. Liposome Res. 5, 605 (1995); (4) L. T. Mimms et al., Biochemistry 20, 833 (1981); (5) L. D. Mayer et al., Biochim. Biophys. Acta 817, 193 (1985); (6) F. Szoka Jr., and D. Papahadjopoulos, Proc. Natl. Acad. Sci. USA 75, 4194 (1978); (7) S. M. Gruner et al., Biochemistry 24, 2833 (1985); (8) L. D. Mayer et al., Biochim. Biophys. Acta 858, 161 (1986); (9) W. R. Perkins et al., Biochim. Biophys. Acta 943, 103 (1988); (10) K. Anzai et al., Biochim. Biophys. Acta 937, 73 (1988).

percentage of unsaturated phospholipids such as DOPC, POPC, and EPC will not form IF-liposomes. Liposome formulations containing a high percentage of fully saturated phosphatidylcholines such as DMPC (di-14:0), DPPC (di-16:0), DSPC (di-18:0), DAPC (di-20:0), and the ether-linked phosphatidylcholine DHPC (di-16:0) should form both ethanol-induced and pressure-induced IF-liposomes. In addition, mixtures of these phosphatidylcholines will form IF-liposomes if the difference in hydrocarbon chain lengths is not too great.[4] The negatively charged phospholipid DPPG (1,2-dipalmitoyl-*sn*-glycero-3-[phospho-rac-(1-glycerol)]) and DPPG–DPPC mixtures will form IF-liposomes.[4] Incorporation of cholesterol directly into the precursor SUVs prevents the formation of IF-liposomes, probably because cholesterol inhibits bilayer interdigitation.[24] High levels of cholesterol incorporation into DPPC IF-liposomes can be obtained, however, by adding cholesterol-containing SUVs to the interdigitated DPPC sheets immediately before raising the sample temperature above the T_m.[4] This modification of the IF process is suitable for introducing other membrane-soluble agents that inhibit interdigitation into the bilayers of IF-liposomes. Thus, even though there are some limits on the choice of phospholipids, basic liposome properties such as surface charge, membrane fluidity, and ion permeability can be controlled and modified in IF-liposome formulations by adjusting the lipid composition.

Liposomes have been used for more than 30 years with great success as model systems for biological membranes. However, most of the standard liposome types are somewhat imperfect model systems for biological membranes. SUV liposomes are unilamellar like typical biological membranes, but for many research applications the small size of SUVs is a significant disadvantage. In addition, the high curvature of SUV bilayers relative to the size of the phospholipids produces forces not found in natural biological membranes. Although MLV liposome suspensions contain a high percentage of relatively large liposomes, MLVs are also a rather crude model biomembrane system. The high level of liposome size heterogeneity, the presence of closely packed bilayers, the low internal volume, and the exclusion of solutes often make MLVs unsatisfactory models for biological membranes.[25] Differences in liposome lamellarity can produce changes in basic thermodynamic properties of phospholipid liposomes. Perkins *et al.*[26] found a solute-induced shift in the T_m of IF-liposomes relative to MLV liposomes of the same phospholipid. This T_m shift appeared to be due to an

[24] H. Komatsu and E. S. Rowe, *Biochemistry* **30,** 2463 (1991).

[25] S. M. Gruner, R. P. Lenk, A. S. Janoff, and M. J. Ostro, *Biochemistry* **24,** 2833 (1985).

[26] W. R. Perkins, X. Li, J. L. Slater, P. A. Harmon, P. L. Ahl, S. R. Minchey, S. M. Gruner, and A. S. Janoff, *Biochim. Biophys. Acta* **1327,** 41 (1997).

osmotic stress across the bilayers of unilamellar vesicles resulting from volume reduction generated on adoption of the ripple phase, that is, the $P_{\beta'}$ phase. Although not as well known as the more conventional liposome types, IF-liposomes are beginning to find some biomembrane research applications. DPPC IF-liposomes were used to examine the electroinsertion of the membrane protein glycophorin.[27] The IF-liposomes used in this study were large enough that glycophorin electroinsertion could be observed directly in individual unilamellar IF-liposomes by fluorescence microscopy. IF-liposomes were also used to examine how liposome size affects *in vivo* circulation lifetime in rats.[28] DSPC IF-liposomes form large aggregates below the T_m of DSPC, dramatically increasing the effective size of the circulating particles. Thus, an accurate estimation of average IF-liposome size during *in vivo* circulation lifetime experiment was difficult. The addition of *n*-glutaryl-1,2-dipalmitoyl-*sn*-glycero-3-phosphoethanolamine (*N*-glutaryl-DPPE) at 10 mol% gave the IF-liposomes a negative surface sufficient to prevent liposome aggregation and allowed accurate determination of the *in vivo* circulation lifetime based on the true average size of the large individual liposomes. The number of biomembrane research applications for IF-liposomes should increase if more investigators realize how simple and easy it is to make ethanol-induced IF-liposomes.

The *in vivo* fate of IF-liposomes after intravenous injection is rapid clearance via uptake by resident phagocytic cells of the reticuloendothelial system (RES) that reside primarily in the liver and spleen.[28] The relatively short circulation lifetime of IF-liposomes limits the application of these liposomes as an intravenous drug delivery vehicle targeted to specific sites within the body, such as tumors.[29] However, IF-liposomes are an excellent vehicle to target therapeutic or diagnostic agents to the macrophages of the RES. The rapid accumulation of IF-liposomes in the phagocytic cells of the liver and spleen has been used to image these organs by x-ray computed tomography (CT) with encapsulated water-soluble radio-opaque compounds such as diatrizoate, iotrolan, ioversol, and ioxaglate.[30–32] These

[27] S. Raffy and J. Teissie, *J. Biol. Chem.* **272**, 25524 (1997).
[28] P. L. Ahl, S. K. Bhatia, P. Meers, P. Roberts, R. Stevens, R. Dause, W. R. Perkins, and A. S. Janoff, *Biochim. Biophys. Acta* **1329**, 370 (1997).
[29] W. R. Perkins, *in* "Liposome Rational Design" (A. S. Janoff, ed.), p. 219. Marcel Dekker, New York, 1999.
[30] A. S. Janoff, S. R. Minchey, W. R. Perkins, L. T. Boni, S. E. Seltzer, D. F. Adams, and M. Blau, *Invest. Radiol.* **26**(Suppl. 1), S167 (1991).
[31] S. R. Minchey, W. R. Perkins, P. L. Ahl, J. L. Slater, S. E. Seltzer, M. Blau, D. F. Adams, and A. S. Janoff, *Biophys. J.* **61**, A447 (1992).
[32] S. E. Seltzer, A. S. Janoff, M. Blau, D. F. Adams, S. R. Minchey, and L. T. Boni, *Invest. Radiol.* **26**(Suppl. 1), S169 (1991).

agents can be encapsulated easily into DSPC IF-liposomes at high iodine-to-lipid ratios and high levels of contrast enhancement in the liver and spleen have been demonstrated in several x-ray CT imaging studies using these agents.[30,32] The passive targeting of IF-liposomes containing antithrombotic agents to blood clots is another possible medical application for IF-liposomes. Streptokinase-encapsulating IF-liposomes have been shown to increase thrombolysis in both nonimmunized rabbits and rabbits immunized against streptokinase.[33] Amikacin- and gentamicin-encapsulating liposomes have been shown to enhance significantly the activity of both aminoglycoside antibiotics toward intracellular mycobacterial infections.[34–36] Passive targeting to resident macrophages of the liver and spleen by IF-liposomes encapsulating these drugs might be a particularly useful treatment for intracellular infections of the RES. Another possible way to use IF-liposomes for drug delivery would be to inject them directly into the target site, rather than injecting them intravenously. The large size of IF-liposomes would tend to keep them localized near the injection site, where they could act as a slow-release drug reservoir.

Summary

IF-liposomes are formed by a unique process that involves fusing small liposomes into interdigitated lipid sheets, using either ethanol or hydrostatic pressure. The interdigitation–fusion method requires liposome formulations with lipids that form the $L_{\beta I}$ phase. Preparing ethanol-induced IF-liposomes is simple and quick. IF-liposomes are particularly well suited for biomembrane research experiments that require large unilamellar liposomes and for liposome drug delivery applications that require a high drug-to-lipid ratio.

Acknowledgment

This work was supported by Elan Pharmaceuticals, Inc.

[33] W. R. Perkins, D. E. Vaughan, S. R. Plavin, W. L. Daley, J. Rauch, L. Lee, and A. S. Janoff, *Thromb. Haemost.* **77,** 1174 (1997).

[34] E. A. Petersen, J. B. Grayson, E. M. Hersh, R. T. Dorr, S. M. Chiang, M. Oka, and R. T. Proffitt, *J. Antimicrob. Chemother.* **38,** 819 (1996).

[35] S. D. Nightingale, S. L. Saletan, C. E. Swenson, A. J. Lawrence, D. A. Watson, F. G. Pilkiewicz, E. G. Silverman, and S. X. Cal, *Antimicrob. Agents Chemother.* **37,** 1869 (1993).

[36] C. E. Swenson, K. A. Stewart, J. L. Hammett, W. E. Fitzsimmons, and R. S. Ginsberg, *Antimicrob. Agents Chemother.* **34,** 235 (1990).

[8]   Freeze-Drying of Liposomes: Theory and Practice

By EWOUD C. A. VAN WINDEN

Introduction

Freeze-drying of liposomes can prevent hydrolysis of the phospholipids and physical degradation of the vesicles during storage. In addition, it may help stabilize the substance that is incorporated in the liposomes. Freeze-drying of a liposome formulation results in an elegant dry cake, which can be reconstituted within seconds to obtain the original dispersion, that is, if the appropriate excipients are used and if suitable freeze-drying conditions are applied. On the other hand, the freeze-drying process itself may induce physical changes of the liposomes, such as a loss of the encapsulated agent and alterations in the vesicle size. The occurrence of such damage is not surprising, because interaction between the hydrophilic phospholipid head groups and water molecules plays a key role in the formation of liposomal bilayers. Thus, removing water from the liposomes by freeze-drying represents an exciting challenge. Moreover, freeze-drying (also called lyophilization) is a time- and energy-consuming process, which certainly requires some expertise in order to avoid its specific pitfalls. Fortunately, excipients, such as disaccharides, have been identified that protect the liposomes during the freeze-drying process (lyoprotectants) and the freeze-drying technique has been described extensively in the literature (see, e.g., Refs. 1–4 for further reading).

In this chapter guidelines are given for optimizing the liposome formulation and the freeze-drying process, and an example of a freeze-drying protocol is described. It is possible to distinguish different types of liposome formulations with respect to freeze-drying: (1) empty liposomes, which are reconstituted with a solution of the compound to be encapsulated,[5–7] (2) liposomes loaded with a compound that is strongly associated

[1] M. J. Pikal, S. Shah, M. L. Roy, and R. Putman, *Int. J. Pharm. Sci.* **60,** 203 (1990).

[2] M. J. Pikal, *BioPharm* **10,** 18 (1990).

[3] D. Essig, R. Oschmann, and W. Schwabe, *in* "Lyophilization." Wissenschaftliche Verlagsgesellschaft, Stuttgart, Germany, 1993.

[4] T. Jenning, "Lyophilization: Introduction and Basic Principles." Interpharm Press, Englewood, CO, 1999; see also www.phase-technologies.com.

[5] Product description of Coatsome EL series, NOF Corporation, Japan, 1995.

[6] W. Zhang, E. C. A. Van Winden, J. A. Bouwstra, and D. J. A. Crommelin, *Cryobiology* **35,** 277 (1997).

[7] D. Peer and R. Margalit, *Arch. Biochem. Biophys.* **383,** 185 (2000).

with the bilayer, and (3) liposomes that contain a water-soluble compound that does not interact with the bilayer. This chapter focuses especially on the freeze-drying of liposomes loaded with a water-soluble compound, because it represents the greatest challenge: both prevention of leakage of encapsulated solutes and preservation of liposome size are required. For liposomes that do not contain a water-soluble compound, formulation and process parameters required to obtain optimal results may be less strict. In practice, many compounds associate with the lipid bilayer but also partially partition into the aqueous phase, which may result in an intermediate type of behavior during freeze-drying. These cases may require special attention for the reason that dilution of such liposome dispersions often results in a reduction of the amount of compound associated with the liposomes.

Freeze-Drying Process

Generally, three phases can be distinguished in the freeze-drying process: (1) freezing, (2) primary drying, and (3) secondary drying. The parameters of each phase depend on the selected formulation and can determine the quality of the final product. Therefore, a basic understanding of the process is indispensable in order to tailor the process parameters to specific needs, and to allow for troubleshooting in case the results are not satisfactory. Below, the effects that process parameters can have on the stability of liposomes are discussed for each of the freeze-drying phases. An overview of parameters that affect the characteristics of the freeze-dried liposome product, as well as suggestions to optimize its quality, is given in Table I.

Freezing Phase

In the freezing phase, cooling of the sample results in the formation of ice crystals. As a result, both the remaining solutes and the liposomes become more and more concentrated ("freeze-concentration"). At this stage, the lyoprotectant forms an amorphous (noncrystalline) matrix in and around the liposomes that prevents fusion processes and protects the liposomes against rupture due to the growth of ice crystals. As the temperature of the freeze-concentrate is lowered and its water content is reduced as a result of the progression of ice formation, the viscosity of the freeze-concentrate increases, until it forms a glass with low molecular mobility. The temperature at which this occurs is called $T_{g'}$, the glass transition temperature of the freeze-concentrated solution, or, alternatively, the collapse temperature (T_c). The $T_{g'}$ is a key parameter for the primary drying process (see below). Generally, the product is cooled to about $5°$ below $T_{g'}$ before the start of the drying phase. The $T_{g'}$ depends on the composition of the

TABLE I

STRATEGIES FOR OPTIMIZATION OF FREEZE-DRYING PROCESS FOR LIPOSOMES

Requirement	Measures	Suggestions
Preservation of liposome size	Add amorphous lyoprotective matrix inside and outside the liposomes.	1. Use disaccharides (sucrose, trehalose, lactose) at >2 g/g phospholipid both inside and outside the liposomes. Avoid the use of crystallizing excipients (e.g., glycine, mannitol).
	Small, unilamellar liposomes	2. Reduce size to 0.1–0.2 μm by homogenization, extrusion, or sonication. Both smaller and larger liposomes show more leakage of encapsulated compounds.
	Freezing after undercooling	3. Cool at 0.5°/min or less. Freezing on a prechilled shelf or in boiling nitrogen results for some liposome types in enhanced leakage of encapsulated compounds.
Retention of encapsulated, water-soluble compound after reconstitution	Rigid bilayers	4. Try the use of hydrogenated phospholipids (e.g., HEPS, HSPC, DPPC). Add cholesterol (e.g., 30 mol%) to the bilayer
	Prevent bilayer rupture by crystal growth (ice, salts, other components), provide substitute for water.	5. See suggestion 1
	Temperature at sublimation front and of dried cake must remain below T_g of the amorphous matrix throughout the freeze-drying process	6. Control the pressure during sublimation of ice at a value corresponding to an ice temperature that is 4° below T_g' of the amorphous matrix. Do not start the secondary drying phase before the completion of ice sublimation. Increase the shelf temperature more slowly.
Easy reconstitution; cake with attractive appearance (no collapse)	Formulate towards a high T_g of the amorphous matrix. Add a bulking agent.	7. Use a disaccharide instead of glucose or sorbitol and minimize the use of NaCl. Minimal content of solids before drying in order to obtain a cake: 2%. Maximal content that allows for drying: ~30%.
Short freeze-dry cycle	Formation of large ice crystals during freezing phase	8. See suggestion 3. One may also apply "annealing" (at, e.g., −10°) of the frozen sample before the start of primary drying (see text).
	Decrease the filling height of the sample.	

(continued)

TABLE I   (*Continued*)

Requirement	Measures	Suggestions
Preservation of labile macromolecules (proteins) ⟶	{ Prevent pH changes during freezing.	9. Avoid the use of sodium phosphate salts, which may cause large pH shifts during freezing.
Chemical and physical stability of dried product ⟶	{ Storage at a temperature well below T_g of the dried cake. Lower residual water content of (amorphous matrix of) the dried cake. Crystalline water does not necessarily contribute to degradation processes. Minimize the oxygen content in the freeze-dried cake.	10. See suggestion 7. Dry to a low residual water content (e.g., <1%) in order to obtain a high T_g. 11. Prolong the secondary drying phase. Do not overheat the shelf during primary drying in order to avoid microcollapse, which slows down secondary drying. Use small filling heights of the samples to be freeze-dried (e.g., not more than 35% of the nominal vial volume). Use dried stoppers (e.g., 3 h at 110°) in order to avoid absorption of water from the rubber by the dried, hygroscopic cake. Release vacuum with nitrogen dry gas before stoppering.

TABLE II
GLASS TRANSITION TEMPERATURES OF MAXIMALLY
FREEZE-CONCENTRATED EXCIPIENTS

Carbohydrate	$T_{g'}$ (°C)
Glucose	−43
Sorbitol	−43
Fructose	−42
Galactose	−42
Sucrose	−32
Maltose	−29
Trehalose	−30
Lactose	−28
Raffinose	−26
Dextran	−14

amorphous freeze-concentrate, which in most cases contains a lyoprotectant and a minimal amount of buffer components. The $T_{g'}$ values of some lyoprotectants are listed in Table II. Liposomes do not contribute to the value of $T_{g'}$. In the case of mixtures of excipients, the $T_{g'}$ is best determined by differential scanning calorimetry (DSC)[8,9] or electrical resistance analysis during thawing.[10] The presence of small molecules such as sodium chloride in the amorphous matrix can strongly decrease the $T_{g'}$ and should be avoided for reasons that are explained below.

The cooling rate of the freezing phase is also an important factor. Quick cooling (on a prechilled shelf or in boiling nitrogen) can result in the formation of dendritic ice crystals that grow from the bottom of the vial to the top of the fluid. Thus, the freeze-concentrate is pushed up, and it finally forms an amorphous layer at the top of the sample in which hardly any ice crystals have been formed. This will result in a crust on top of the cake with only a small amount of pores, which will slow the escape of water vapor from the sample. Therefore, slow cooling (e.g., 0.5°/min) is preferred, because this allows for undercooling of the sample: the temperature reaches values below 0° without immediate ice formation. Once freezing occurs, the whole sample freezes nearly instantaneously, resulting in similar crystal sizes throughout the product. Such pores reduce the resistance of the cake to water vapor during sublimation.[11] Moreover, slow cooling has been shown

[8] L. M. Her and S. L. Nail, *Pharm. Res.* **11,** 54 (1994).
[9] E. C. A. Van Winden, H. Talsma, and D. J. A. Crommelin, *J. Pharm. Sci.* **87,** 231 (1998).
[10] L. M. Her, R. P. Jefferis, L. A. Gatlin, B. Braxton, and S. L. Nail, *Pharm. Res.* **11,** 1023 (1994).
[11] J. A. Searles, J. F. Carpenter, and T. W. Randolph, *J. Pharm. Sci.* **90,** 872 (2001).

to minimize the leakage of contents from liposomes after freeze-drying and rehydration, depending on the liposome type, as is described later.[12] The reasons behind this phenomenon have not been clarified yet.

Crystallization of the ice can be optimized further by "annealing" of the frozen samples: the samples are heated to, for example, $-10°$ (above $T_{g'}$) and are cooled again before the start of the freeze-drying process. The energy that becomes available for the frozen matrix may induce further crystallization of unfrozen water in the amorphous matrix ("devitrification"). In addition, small crystals may recrystallize into larger ones. The result of annealing is often an increase in sublimation rate during primary drying.[11] In addition, liposomes may be better protected against freeze-drying stress.[12] However, only limited data on the effect of annealing on liposomes are available so far, and the effect of annealing may depend on the specific formulation (e.g., excipients, lipid composition, and vesicle size).

Primary Drying

During primary drying the ice crystals are sublimated, resulting in the porous cake structure of the freeze-concentrated matrix. As long as the product temperature or, more precisely, the temperature at the sublimation front (T_{front}), is maintained at about $4°$ below $T_{g'}$, degradation of the liposomes is minimized and ice crystals can be removed by sublimation without collapse of the cake structure. However, if T_{front} rises above $T_{g'}$, (partial) collapse of the cake occurs, which results in a reduced surface area of the dried matrix. As a consequence, the final residual water content after drying may increase, as well as the time required for reconstitution, and the liposomes may fuse and lose part of their contents on rehydration.

How can T_{front} be controlled during primary drying? The product temperature at the sublimation front is determined by local pressure, according to the ice vapor pressure–temperature curve (see Table III). For this reason, the pressure in the freeze-drying chamber should be controlled well. The set point of the pressure is chosen at a value that corresponds to an ice temperature of about $4°$ below $T_{g'}$. The shelf temperature should be higher than the ice temperature in order to transfer the heat to the sublimation front that is required for drying. The driving force of the drying process is the difference in temperature and, therefore, difference in water vapor pressure, between the condenser of the freeze-dryer and the sublimation front. The condenser temperatures of most commercial freeze-dryers reach $-55°$ or lower.

[12] E. C. A. Van Winden, W. Zhang, and D. J. A. Crommelin, *Pharm. Res.* **14,** 1151 (1997).

TABLE III
ICE TEMPERATURE AS FUNCTION OF VAPOR PRESSURE

Ice temperature (°C)	Pressure (matm)
−12	2.051
−16	1.438
−20	0.996
−24	0.681
−26	0.560
−28	0.459
−30	0.375
−32	0.305
−34	0.247
−36	0.200
−38	0.160
−40	0.128
−42	0.102
−44	0.081

A limiting factor in the drying process can be the heat transfer from the shelf to the sublimation front, which occurs via conduction by glass and ice, as well as by gas molecules.[2,13] Therefore, more heat is available for sublimation at the ice front at a higher chamber pressure. The set point for the chamber pressure is, however, restrained by the $T_{g'}$ of the product, because increasing the pressure set point should not increase the ice temperature to a value close to $T_{g'}$ or higher. Neither can the shelf temperature be increased without limitation, for the following reasons: first, the rise in temperature of the frozen sample between the shelf and the sublimation front should not impair the product stability. Second, more energy becomes available for sublimation as the ice front approaches the bottom of the vial. On the other hand, the increased thickness of the dried cake represents a higher resistance for water vapor and may hamper the evaporation. As a result, the local pressure at the sublimation front inside the cake may become higher than the chamber pressure and, therefore, T_{front} may exceed $T_{g'}$, which leads to collapse.

From the above it will be evident that the geometry of the vial and the filling height play a crucial role in the drying process. Process development must, therefore, be performed with the vials and sample volume selected for the final product. A maximal filling height of 30% of the vial is recommended in order to prevent excessively long drying processes. As

[13] H. Ybema, L. Kolkman-Roodbeen, M. P. Te Booy, and H. Vromans, *Pharm. Res.* **12,** 1260 (1995).

mentioned above, the cooling protocol may have a profound effect on the sublimation rate. Even the choice of freeze-dry stoppers or their incorrect placement may affect the outcome of the freeze-dry run in certain cases. In conclusion, careful control of chamber pressure and shelf temperature, as well as knowledge of the sample $T_{g'}$, are prerequisites for successful freeze-drying.

Secondary Drying

In the secondary drying phase, the residual water content of the amorphous matrix is further decreased. After all the ice has disappeared and the product temperature has approached the value of the shelf temperature, the secondary drying phase can be initiated. For this reason, temperature probes provide useful information during the freeze-dry run. In the secondary drying phase, the pressure is reduced to a minimal value and the shelf temperature can be gradually increased to, for example, $25°$. Reduction of the water content of the cake also occurs during primary drying, but lower values ($<1\%$) can be achieved by adjusting the conditions. Reduction of the water content increases the glass transition temperature (T_g) of the cake, and this allows for a gradual increase in shelf temperature. The shelf temperature is a key parameter for obtaining a low residual water content. Generally, a freeze-dried product is likely to be stable at 20 to $30°$ below T_g and, therefore, reduction of the residual water content is of utmost importance. It is worthwhile to mention that the T_g of the amorphous cake is not the only parameter that affects the stability of the freeze-dried liposomes. Heating freeze-dried dipalmitoyl phosphatidylcholine (DPPC) liposomes to the melting transition of the bilayers in the dried state for only 30 min has also been shown to result in leakage of the liposomal contents on rehydration, even though the glass transition temperature had not been reached.[14] Further, the T_g of the intraliposomal matrix may differ from the T_g of the extraliposomal matrix because of the presence of a high concentration of encapsulated compound. Chemical degradation of the encapsulated compound may, in this case, account for degradation processes observed well below $T_{g'}$ of the extraliposomal matrix.[15] Although physical processes that can occur in a cake of freeze-dried liposomes are rather complex, low water contents are generally expected to enhance the stability of the dried product.

Vials can be closed under vacuum or after (partially) releasing the vacuum with nitrogen gas or argon. Filling the vials with air results in

[14] E. C. A. Van Winden and D. J. A. Crommelin, *J. Control. Release* **58,** 69 (1999).
[15] E. C. A. Van Winden and D. J. A. Crommelin, *Eur. J. Pharm. Biopharm.* **43,** 295 (1997).

oxidation of unsaturated phospholipids during storage (N. J. Zuidam, personal communication).

Selection of Liposome Type

The optimal liposome size for high retention of a water-soluble compound after freeze-drying is 0.1–0.2 μm.[16,17] Possible explanations for the higher leakage observed for smaller liposomes, or larger, multilamellar liposomes are discussed elsewhere.[18] High retention values have been reported for a large variety of bilayer compositions.[19] Some inconsistencies between different studies exist, and these could be due to differences in process parameters or choice of lyoprotectants and other excipients. Rigid liposomes are more vulnerable to high cooling rates than liposomes with flexible bilayers, but give a high retention (70–90%) of the encapsulated hydrophilic marker when applying a cooling rate of about 0.5°/min.[12] This effect of the cooling rate has been observed in samples with sucrose, trehalose, as well as glucose as a lyoprotectant. Rigid liposomes can be composed of phospholipids with saturated acyl chains such as DPPC or hydrogenated egg phosphatidylcholine (HEPC).

For liposomes containing unsaturated egg phosphatidylcholine (EPC) addition of the negatively charged phospholipid phosphatidylglycerol (PG) has been shown to decrease the retention, whereas addition of phosphatidylserine (PS) increases the retention, of a water-soluble encapsulated marker after freeze-drying and rehydration.[16] On the other hand, an equally high retention of an encapsulated water-soluble marker, carboxyfluorescein (CF), in DPPC–dipalmitoylphosphatidylglycerol (DPPG) (molar ratio, 10:1) and DPPC liposomes has been reported[12] and, therefore, the precise effects of addition of the negatively charged PG to the bilayer are not clear. Addition of cholesterol (e.g., 30 mol% of the total lipid content) may increase the resistance of liposomal bilayers to freeze-drying stress. Dioleoylphosphatidylcholine (DOPC) liposomes lose almost all their contents when freeze-dried in the presence of maltose, in contrast to liposomes made of DPPC and EPC.[19]

[16] J. H. Crowe and L. M. Crowe, in "Liposome Technology" (G. Gregoriadis, ed.), Vol. I. CRC Press, Boca Raton, FL, 1993.

[17] J. H. Crowe and L. M. Crowe, *Biochim. Biophys. Acta* **939,** 327 (1988).

[18] E. C. A. Van Winden, N. J. Zuidam, and D. J. A. Crommelin, in "Medical Applications of Liposomes" (D. D. Lasic and D. Papahadjopoulos, eds.), p. 567. Elsevier Sciences, Amsterdam, 1998.

[19] H. Komatsu, H. Saito, S. Okada, M. Tanaka, M. Egashira, and T. Handa, *Chem. Phys. Lipids* **113,** 29 (2001).

In summary, the bilayer composition can determine the resistance of the liposomes to freeze-drying stress, but it is difficult to extract general rules from the literature because of the many other parameters that are involved, such as process conditions, choice of lyoprotectant, and vesicle size.

Lyoprotectants and Other Excipients

Lyoprotectants protect liposomes by (1) preventing fusion of liposomes, (2) preventing the rupture of bilayers by ice crystals, and (3) maintaining the integrity of the bilayers in the absence of water. To do so, the lyoprotectants must form an amorphous, glassy matrix in and around the liposomes. Interaction between the lyoprotectant and the phospholipid head groups is considered especially important for preventing leakage during rehydration of liposomes that have a liquid-crystalline bilayer in the hydrated state at ambient temperature.[20] More recently, a different mechanism was proposed by which sugars may reduce the stress on bilayers caused by drying. Bilayers dried without lycoprotectant approach each other closely and the suction of water by the phospholipid causes a lateral comprehensive stress. The mere presence of a glass between adjacent bilayers during drying provides spacing between the bilayers, and in combination with the attraction of water by the sugars (osmotic properties) the compressive stress on the bilayers is reduced. For more literature on the mechanism of lyoprotection of liposomes the reader is referred to Refs. 12 and 20–22.

The preferred lyoprotectants are disaccharides such as sucrose, trehalose, maltose, and lactose. Trehalose, maltose, and lactose have a higher T_g (100–115°) in the dried state (<0.5% residual water) than sucrose (~70°), which may be an advantage for the storage stability of the dried product. Lyoprotectants provide optimal protection of the liposomes if present at a ratio of at least 2 g/g phospholipid. In addition, the lyoprotectant should be present preferably both inside and outside the liposomes, at equimolar concentrations.[17] The lyoprotectant also serves as a bulking agent (cake former). Generally, freeze-dried cakes have a good consistency and appearance if the original dispersion has a total solid concentration of at least 2.5% (w/v). A total solute concentration of more than about 30% may hinder the drying process. Dextran is a polymer that also forms an amorphous matrix on freeze-drying. However, it fails to prevent leakage of a water-soluble compound from EPC liposomes on rehydration, possibly

[20] J. H. Crowe, S. B. Leslie, and L. M. Crowe, *Cryobiology* **31**, 355 (1994).
[21] J. Wolfe and G. Bryant, *Cryobiology* **39**, 103 (1999).
[22] K. L. Koster, Y. P. Lei, M. Anderson, S. Martin, and G. Bryant, *Biophys. J.* **78**, 1932 (2000).

because it does not interact with the phospholipid head groups.[20] Recently, an alternative explanation for the lack of protection of bilayers by dextran was provided by Koster et al.,[23] who demonstrated that dextran with a molecular size of 1000 or lower was able to suppress the rises of T_m upon dehydration, whereas larger dextrans (5000 and 12000) did not have such an effect. The authors concluded that larger molecules are excluded from the interlamellar spece of bilayer systems during dehydration and, therefore do not reduce the lateral compression that is caused by drying. The use of crystallizing excipients, such as glycine and mannitol, is not recommended for the freeze-drying of liposomes. A (partial) crystalline excipient can facilitate the primary drying process and improve the appearance of the freeze-dried cake. However, their successful use for the freeze-drying of liposomes loaded with water-soluble compounds has not been reported so far, and the lack of an amorphous phase in and around the liposomes may damage the bilayers.

The use of excipients with a low molecular weight, such as sodium chloride, glucose, and amino acids, should be minimized. If such excipients remain amorphous in the sugar matrix they will lower the $T_{g'}$ of the freeze-concentrate and the T_g of the freeze-dried cake. A low $T_{g'}$ requires freeze-drying at a low product temperature, which is an inefficient process. A decrease in product temperature of 5° may even double the time required for primary drying. In addition, a low T_g may reduce the stability of the dried cake (see above).

Freeze-Drying of Liposomes Loaded with Non–Water-Soluble Compounds

If the compound incorporated in the liposomes associates strongly with the bilayer, no risk of leakage exists and some of the required formulation characteristics or process parameters may be less critical. For instance, the vesicle size may be larger than 0.2 μm and quick cooling may give the same result as slow cooling. Reports have appeared on the freeze-drying of complexes that are designed for their use in gene therapy.[23–25] From these studies it is clear that lipid–DNA complexes can also be freeze-dried without causing major alterations of their physical characteristics and transfection efficiencies. In one case it was demonstrated that the freeze-dried product was stable at room temperature for 8 weeks without significant changes in the product characteristics.[23]

[23] K. L. Koster, K. J. Maddocks, and G. Bryant, *Eur. Biophys. J.*. **32**, 94–105 (2003).
[24] R. Cortesi, E. Esposito, and C. Nastruzzi, *Antisense Nucleic Acid Drug Dev*. **10**, 205 (2000).
[25] M. C. Molina, S. D. Allison, and T. J. Anchordoquy, *J. Pharm. Sci*. **90**, 1445 (2001).

Freeze-Drying Protocol

Materials

Vials: 10 R freeze-dry vials (10 ml)

Filling height: 2 ml

Liposomes: DPPC–DPPG (10:1, mol/mol); average size, 120 nm

Intraliposomal composition: 10% sucrose, 5 mM CF, 30 mM sodium citrate (pH 6.5)

Extraliposomal medium: 10% sucrose, 30 mM sodium citrate (pH 6.5) ($T_{g'} = -33°$)

Stoppers: For example, type V 9172 DZ, FM 257/5 (Helvoet Pharma, Alken, Belgium). The stoppers should be dried for 3 h at 110° before use.

Freeze–Drying Protocol

1. The semistoppered vials are loaded on the shelf.
2. The shelf is cooled to 0° (as fast as possible to save time).
3. The shelf is cooled at 0.5°/min to −40°.
4. The samples are maintained at −40° for 15 min.
5. The pressure is reduced to 0.12–0.16 matm.
6. The shelf temperature is increased to −25°.
7. The shelf temperature is maintained at −25° for 30 h.
8. The shelf is heated to 25° at 5°/min.
9. The vacuum is reduced to a minimal value for 6 h.
10. The vacuum is released with nitrogen gas 5.0 (high purity).
11. The vials are stoppered.
12. The freeze-dryer is opened and the vials are removed.
13. The vials are sealed.

[9] Preparation of Sterile Liposomes by Proliposome–Liposome Method

By Jaroslav Turánek, Andrea Kašná, Dana Záluská and Jiří Neča

Introduction

Liposomes are microscopic membrane-like spherical structures consisting of one or more concentric lipid bilayers enclosing aqueous compartments, and have become a useful tool and model in various areas of science. Commercial liposomal drugs are already available.[1] Liposomes were adopted by numerous researchers as the vehicle of choice for drug and vaccine delivery and targeting.[2,3]

Liposomes are classified in terms of the number of bilayers enclosing the sequestered aqueous volume into unilamellar, oligolamellar, and multilamellar vesicles. Unilamellar vesicles can be divided into small unilamellar vesicles (SUVs) with a large curvature, and large unilamellar vesicles (LUVs) with a low curvature, and hence with properties similar to those of a flat surface. Multilamellar vesicles (MLVs) are liposomes representing a heterogeneous group in terms of size and morphology.[4] Lipid composition, size, and morphology are variables determining the fate of liposomes in the biological milieu; therefore selection of a suitable method of preparation of liposomal drugs is of importance for both design of animal experiments and future successful marketing of the product.

A variety of procedures for the preparation of different types of liposomes have been developed and summarized in several reviews and monographs, including this series.[5,6] These methods have been classified for convenience into three categories: (1) mechanical dispersion methods, such as hand shaking or vortexing, sonication, and high-pressure homogenization, (2) detergent-solubilizing dispersion methods including solubilized lecithin dispersion with sodium cholate or octylglucoside, and (3) solvent dispersion methods such as ethanol injection, ether infusion, and

[1] G. Gregoriadis, *Trends Biotechnol.* **13,** 527 (1995).

[2] G. Lopez-Berestein and I. J. Fidler, "Liposomes in the Therapy of Infectious Diseases and Cancer." A. R. Liss, New York, 1989.

[3] C. R. Alving, *J. Immunol. Methods* **140,** 1 (1991).

[4] P. R. Cullis, M. J. Hope, M. B. Bally, T. P. Madden, and L. D. Mayer, *in* "Liposomes: From Biophysics to Therapeutics" (M. J. Ostro, ed.), p. 39. Marcel Dekker, New York, 1987.

[5] M. C. Woodle and D. Papahadjopoulos, *Methods Enzymol.* **171,** 193 (1988).

[6] G. Gregoriadis, "Liposome Technology." CRC Press, Boca Raton, FL, 1992.

reversed-phase evaporation. These primary processes can be linked with the secondary processes, such as high-pressure homogenization or extrusion through polycarbonate filters with various pore sizes, which are easy ways of obtaining oligolamellar or unilamellar liposomes with various size distributions.[4,7–11]

Ethanol solvent dispersion (ethanol injection method) has some drawbacks, including low encapsulation efficiency and restriction of lipid composition owing to limitations in the solubility of some lipids in ethanol.

The proliposome–liposome method is based on proliposomal preparation of hydrated stacked bilayer sheets in a water–ethanol solution. Generally, the organization of a lipid–ethanol–water mixture can be described in terms of a three-phase diagram that can be divided into the following principal areas: lipid dissolved in aqueous ethanol, hydrated bilayers suspended in aqueous ethanol, and a liposomal area. Spontaneous formation of liposomal suspensions is accomplished by addition of excess aqueous phase to a lipid mixture.

This technique is simple and practical and is characterized by an extremely high entrapment efficiency, when compared with other methods based on passive entrapment. The technique is suitable for encapsulation of a wide range of drugs with various water and alcohol solubilities.[12,13]

In this chapter we describe a stirred thermostatted cell and its link-up with a liquid delivery system for the rapid production of multilamellar liposomes by the proliposome–liposome method, which is based on the conversion of the initial proliposome preparation into a liposome dispersion by dilution under strictly controlled conditions. The cell has been designed for laboratory-scale preparation of liposomes (300–1000 mg of phospholipid per run) in a procedure taking less than 90 min and can be readily scaled up and linked with secondary processing methods, such as extrusion through polycarbonate filters for the preparation of oligolamellar and unilamellar liposomes.

The cell is a prospective tool for modeling the processes of industrial-scale preparation of liposomes.

[7] M. J. Hope, M. B. Bally, G. Webb, and P. R. Cullis, *Biochim. Biophys. Acta* **812**, 55 (1985).
[8] J. Turánek, *Anal. Biochem.* **218**, 352 (1994).
[9] N. Berger, A. Sachse, J. Bender, R. Shubert, and M. Brandl, *Int. J. Pharm.* **223**, 55 (2001).
[10] R. Barnadas-Rodríguez and M. Sabés *Int. J. Pharm.* **213**, 175 (2001).
[11] T. Schneider, A. Sachse, G. Rößling, and M. Brandl, *Int. J. Pharm.* **117**, 1 (1995).
[12] S. Perrett, M. Golding, and P. Williams, *J. Pharm. Pharmacol.* **43**, 154 (1991).
[13] J. Turánek, D. Záluská, and J. Neča, *Anal. Biochem.* **249**, 131 (1997).

Experimental Procedures

Stirred Cell and Its Link-up with Thermostat and Delivery Pump

A schematic drawing and the actual appearance of a disassembled cell are shown in Fig. 1 and 2A, respectively. The acrylic body of the thermostatic jacket has been designed to accept a centrifuge tube (50 ml: Corning, Corning, NY) and two O-rings are used to seal the tube in the thermostatic chamber. The diluting solution is delivered through a Teflon capillary (1.8 mm o.d. × 0.5 mm i.d.), and a sterile filter (0.22 μm; Millipore, Prague, Czech Republic) removes residual particles and microorganisms to maintain the preparation sterile. A Teflon body (capillary reinforcement) protects the capillary from unwanted bending (Fig. 2B). A magnetic stirrer hangs loosely on the flanged end of the capillary. The stirrer, with turbine-like blades, is shaped to reach the conical part of the tube near its bottom. The blades are arranged to lift the suspension up from the bottom and maintain homogeneity, especially during the first step of dilution when the proliposome mixture is viscous. The body of the stirrer is made of Teflon and pieces of Teflon-coated ferrite are inserted into its arms (Fig. 2C). A polypropylene frit closes the pressure-compensation hole and protects the interior from airborne contamination. When a run is finished, the head of the cell is removed and the tube is closed with its original lid.

The actual appearance of linking of the cell with a programmable system for delivery of the diluting solution is shown in Fig. 3. A schematic drawing of the arrangement is presented in Fig. 4. We use a fast protein liquid chromatography (FPLC) system (Pharmacia, Uppsala, Sweden) or a peristaltic pump for accurate flow rate programming. The fraction collector (Frac-100; Pharmacia) connected with the peristaltic pump (P1; Pharmacia) or another adequate system can be used to ensure a defined flow rate even if the two-step dilution method is used, and to stop the pump at the end of the run. The thermostat (Julabo U3; Julabo, Seelbach, Germany) is used to maintain constant temperature during the procedure.

Preparation of Proliposome Mixture and Its Conversion into Liposomes

The egg yolk phosphatidylcholine preparation Lipoid E 80 (Lipoid, Ludwigshafen, Germany) is used in our experiments. A proliposome mixture is prepared by a modification of the method described by Perrett et al.[12] Briefly, the lipids (300 mg of solid yellowish grains) are dissolved in pure warm ethanol (240 mg, 50°) and cooled to room temperature. Water phase is added at a lipid:ethanol:water ratio of 10:8:X (w/w/w). Tris-HCl (20 mM, pH 7.2) is used as the water phase; the value $X = 20$ is

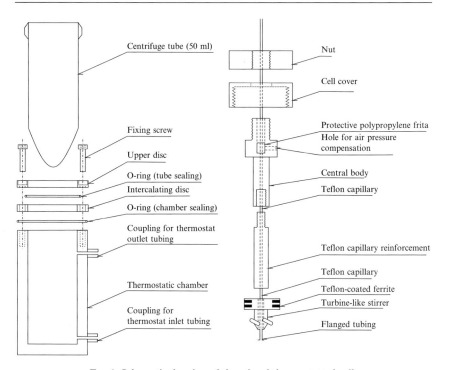

FIG. 1. Schematic drawing of the stirred thermostatted cell.

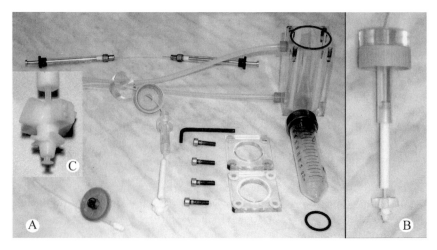

FIG. 2. A disassembled cell (A), assembled cell head with stirrer (B), and detail of stirrer (C).

F IG. 3. Entire system for the preparation of liposomes by the proliposome–liposome method.

used in most experiments (600 mg of water phase). The compounds to be entrapped are dissolved in the water phase and added at this step. Poorly water-soluble compounds are preferably dissolved in ethanol together with phospholipids. Thorough mixing of an ethanol solution of phospholipids with the water phase is achieved by the use of two high-pressure glass syringes, parts of the hand-operated miniextruder supplied by Avestin (Toronto, ON, Canada) or Avanti Polar Lipids (Alabaster, AL) linked with short Teflon capillary tubing (Fig. 2). The water phase with compounds to be entrapped is injected quickly from one syringe into the second syringe containing the ethanol solution of phospholipids. This process is repeated rapidly several times to get a homogeneous preparation. The opaque viscous mixture is heated to $60°$ (transformation temperature) for 10 min in a sterile Eppendorf tube and then allowed to cool to room temperature, yielding a proliposome mixture. The proliposome mixture is transferred into the stirred cell and converted into a liposome suspension by continuous dilution with the water phase (total volume of 39 ml) at a defined flow rate and temperature. The two-step dilution, which is the most effective dilution mode with respect to encapsulation efficiency and

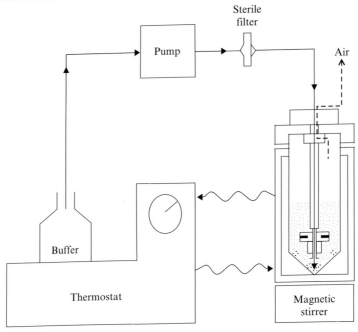

FIG. 4. Schematic drawing of link-up of the cell thermostat and pump. Operation is described in text.

duration of the procedure, is accomplished by continual addition of 3 ml of the water phase at a flow rate of 100 μl/min in the first step and 36 ml of the water phase at 1000 μl/min in the second step.[13]

Assay of Encapsulation Efficiency for 5,6-Carboxyfluorescein

5,6-Carboxyfluorescein (CF; Molecular Probes, Eugene, OR) is quantified spectrofluorimetrically in a luminescence spectrometer (LS 50B; PerkinElmer, Prague, Czech Republic) at 470 nm (excitation) and 520 nm (emission). Free 5,6-carboxyfluorescein is separated from liposome-entrapped CF by centrifugation (45,000g, 20 min, 4°). Liposomes are solubilized with Triton X-100 [final concentration, 1% (w/v)] to release the entrapped drug for assay. The entrapment efficiency is calculated by the equation

$$\text{Percent entrapment} = \left(\frac{C}{A}\right) \times 100$$

and the validity of the calculation is checked by the equation

$$A = B + C$$

where A is total (i.e., free and encapsulated) CF in the liposomal suspension, B is free (nonencapsulated) CF, and C is encapsulated CF.

Effect of Proliposome Mixture Composition on Entrapment Efficiency

The entrapment efficiency of the liposomes formed after dilution of proliposome mixtures to a final volume of 40 ml is plotted as a function of the weight of the aqueous phase present in the initial proliposome mixtures. A typical plot showing the efficiency of entrapment of CF by Lipoid 80 E liposomes is presented in Fig. 5. The region of maximum entrapment efficiency has been mapped in detail. A high entrapment efficiency is obtained with a proliposome lipid:ethanol:water composition of 10:8:15–23 (w/w/w). The proliposome lipid:ethanol:water composition of 10:8:20 (w/w/w) has been used for further studies. Liposomes prepared from a proliposome mixture with this composition yielded an entrapment efficiency of 81 ± 2% for CF (five separate preparations). A negatively stained electron micrograph of liposomes prepared by the proliposome–liposome method is shown in Fig. 6A. Heterogeneity of the liposomal preparation in terms of size, lamellarity and morphology can be seen clearly. Multicentric liposomes are also present.

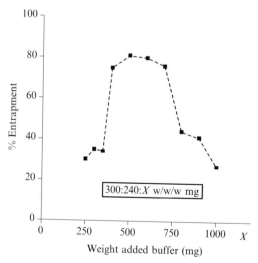

FIG. 5. Plot of entrapment percentage of carboxyfluorescein as a function of weight of buffer (X) incorporated in the initial proliposome mixture [phospholipid:ethanol:buffer = 300:240:X (w/w/w, mg)]. For details see Experimental Procedures.

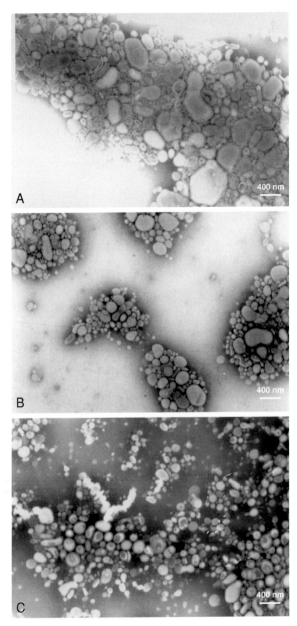

FIG. 6. Electron micrographs of negatively stained liposomes prepared by the proliposome–liposome method. (A) Nonextruded proliposomes; (B) proliposomes extruded through 0.4-μm filters; (C) proliposomes extruded through 0.2-μm filters. Liposomes are stained with 3% ammonium molybdate solution.

Assay of Encapsulation Efficiency for Tested Drugs

The synthetic immunomodulators adamantylamide dipeptide (AdDP) (Lachema, Brno, Czech Republic), muramyl dipeptide (MDP), and β-D-GlcNAc-norMurNAc-L-Abu-D-isoGln (DDD) (obtained by courtesy of M. Ledvina, Institute of Organic Chemistry and Biochemistry, Prague, Czech Republic) is assayed with a high-performance liquid chromatography (HPLC) chromatographic system (Waters, Milford, MA), including a Waters 600 gradient pump, a Waters 717 Plus autoinjector, a Waters 996 diode array detector, a Waters 474 fluorescence detector, and the program Millennium 2010. All the analyses are run on a Waters NOVA-PAC C_{18} column (4 μm, 150 × 3.9 mm; precolumn, 20 × 3.9 mm). Methanol–water (0.1% trifluoroacetic acid, TFA) is used as the mobile phase for the analysis of AdDP (ratio, 45:55) and MDP and DDD (ratio, 10:90). The immunomodulators are detected at 203 nm and the obtained ultraviolet (UV) spectra are matched against the spectra of standards. DDD yields two anomer peaks in the chromatogram, which are both used for the calculation.

The antibiotics gentamicin and neomycin (Sigma, Prague, Czech Republic) are analyzed by precolumn derivatization with *o*-phthalaldehyde and detected by a fluorescence detector set at 340 nm (excitation) and 455 nm (emission).[14,15] The photosensitizer meso-tetra-(*p*-sulfophenyl)-porphine (TPPS$_4$) (kindly provided by Lachema Company) is assayed spectrophotometrically at 410 nm, using the standard addition method. Total masses of free and entrapped drugs are determined and used for the calculation of encapsulation efficiency. Free drugs are separated by centrifugation (45,000g, 20 min, 4°). The liposomes are solubilized with Triton X-100 [final concentration, 1% (w/v)] to release the entrapped drugs for the assay.

Comparison of Entrapment Efficiency for Selected Drugs of Frozen-and-Thawed MLVs and Liposomes Prepared by Proliposome–Liposome Method

Preparation of Frozen-and-Thawed Multilamellar Vesicles. The lipid mixture, dissolved in chloroform (200 mg in 6 ml), is deposited onto the wall of a round-bottom flask (250 ml) by removal of the solvent in a rotary evaporator (40°, 4 h). The dried lipid film is then hydrated with the aqueous phase [150 m*M* NaCl, 20 m*M* HEPES (pH 7.2) previously filtered through a 0.22-μm pore size filter] under continuous mixing on a

[14] V. G. Agarwal, *J. Liq. Chromatogr.* **12**, 3265 (1989).
[15] B. Shaikh, J. Jackson, and G. Guyer *J. Chromatogr.* **571**, 189 (1991).

TABLE I

COMPARISON OF ENTRAPMENT EFFICIENCIES OF FTMLV AND PL LIPOSOMES FOR
SELECTED DRUGS

Drug to be entrapped (mg/300 mg phospholipid)		Percent entrapment			
		Proliposome–liposome		FTMLV	
Drug	Amount	Mean	SD	Mean	SD
A: AdDP	2.28	87[a]	2.2	78	3.2
B: MDP	1.56	62[a]	2.8	36	4.7
C: DDD	1	85[a]	2.8	32	4.1
D: Neomycin	30	65[a]	1.7	36	3.9
E: Gentamicin	30	69[a]	2.2	52	4.2
F: 6-Carboxyfluorescein	1	81[a]	1.5	52	3.9
G: TPPS$_4$	1.3	65	1.5	62	4.0

Abbreviations: FTMLV, Frozen-and-thawed multilamellar vesicle; PL,
proliposome–liposome.
[a] Significant difference from the corresponding results obtained by the FTMLV method
($p < 0.001$, based on t tests).

mechanical reciprocal shaker for 2 h. The frozen-and-thawed MLV
system[7] is obtained by freezing the MLVs in liquid nitrogen and thawing
them in a 30° water bath, and repeating the cycle five times.

The entrapment efficiencies for synthetic immunomodulators and
glycosidic antibiotics are summarized in Table I. Both types of liposomes
show a high entrapment efficiency for the hydrophobic AdDP. The
hydrophilic MDP and DDD are entrapped with a similar efficiency by
frozen-and-thawed MLV (FTMLV) liposomes. Compared with AdDP or
DDD, the entrapment efficiency of PL liposomes for MDP is lower by
approximately 20%. Neomycin and gentamicin are entrapped with similar
efficiencies by PL liposomes, but entrapment efficiencies of FTMLV
liposomes for the two antibiotics are different. TPPS$_4$ is entrapped with
the same efficiency by any of the liposome types.

Determination of Captured Volume. Captured volume, which is defined
as the volume of the entrapped discontinuous aqueous phase per mass
unit of the lipid phase and is expressed in microliters per micromole of
phospholipid, is determined as described by Armengol and Estelrich[16] with
a minor modification. Briefly, the concentration of CF in the proliposome
mixture and in the diluting buffer is 500 μM. Nonentrapped CF is

[16] X. Armengol and J. Estelrich, *J. Microencapsulation* **12**, 525 (1995).

separated from the liposomes by centrifugation (45,000g, 20 min, 4°) and the liposomes are washed twice and resuspended in CF-free buffer to obtain a concentration of 1 mg of phospholipid per milliliter. An aliquot of the liposomes is solubilized with Triton X-100 [final concentration, 1% (w/v)] and the concentration of CF is determined spectrofluorimetrically, using a calibration curve of a CF standard dilution series covering a 10–1000 nM range for CF.

The captured volume is calculated by the equation

$$V_e = \frac{M}{CL}$$

where V_e is the entrapped volume (μl/μmol), M is the total mass of CF in the washed liposome preparation, C is the initial concentration of CF in the diluting buffer and proliposome–liposome mixture, and L is the total mass of lipid in the purified preparation.

Entrapped volumes of frozen-thawed MLVs prepared by hydration of a lipid film and of liposomes prepared by the proliposome–liposome method (without extrusion) are compared. The values of 1.83 ± 0.07 and 1.26 ± 0.06 μl/μmol phospholipid have been found for the former and the latter, respectively.

Electron Microscopy. Electron micrographs are made by the negative staining technique (3% ammonium molybdate solution). A transmission electron microscope (BS 500; Tesla, Brno, Czech Republic) is used. The final magnification of the micrograph prints is × 25,200. The micrographs of some liposomal preparations are presented in Fig. 6.

ζ Potential Measurement and Particle Size Analysis by Photon Correlation Spectroscopy. Light-scattering analyses of liposomal preparations are done with a photon correlation spectrometer (Zetasizer 3000; Malvern Instruments, Malvern, UK). Samples of liposomal preparations (phospholipid, 0.7 mg/ml) are analyzed at 25.0°. ζ potential measurement and a size distribution analysis of the data are done according to the instrument manual.

The ζ potentials of the liposomes prepared from Lipoid E 80 are −18 ± 2.3 mV (20 mM Tris-HCl, pH 7.2) and −4.3 ± 1.7 mV (phosphate-buffered saline).

Effect of Transformation Temperature on Entrapment Efficiency. The physical state of the proliposome mixtures is strongly temperature dependent and the thermal cycle used for the transformation of the proliposome mixture is decisive for the resulting encapsulation efficiency. It should be noted that the transition temperature for egg yolk lecithin membranes is within the range −15 to −7°, and hence the transition from the solid (gel) phase to the liquid-crystal phase does not affect the process

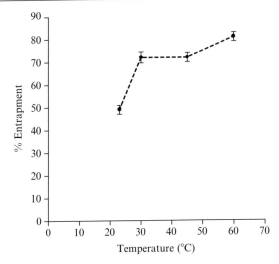

FIG. 7. Effect of transformation temperature on entrapment percentage of carboxyfluo-rescein in a two-step dilution of a proliposome mixture. For details see Experimental Procedures.

of liposome formation at the temperatures used in our experiments. The solubility of lipids in aqueous ethanol increases markedly with rising temperature. This strong temperature dependence is an important feature of the proliposome system. Mixtures at the boundary of the proliposome region can convert from their original proliposome form into liposomes on cooling.

The break point has been found to be 30°. Below 30° there is a steep decrease in entrapment efficiency for CF. The entrapment efficiency of 73% remains constant in the range of 30–45° and increases slowly to 82% when the temperature rises to 60°. These results are in good accordance with the temperature-dependent increase in the isotropic component in ^{31}P nuclear magnetic resonance (^{31}P NMR),[12] reflecting an increase in non-bilayer-organized molecules of phospholipids.[12] The dependence of entrapment efficiency on the transformation temperature is shown in Fig. 7.

Statistical Analyses

An unpaired two-tail *t* test is used to compare the entrapment efficiency for PL liposomes versus FTMLVs. Statistical software includes Prism, version 2.00 (GraphPad Software, San Diego, CA).

Extrusion of Proliposomes

The proliposomes can be extruded with a hand-operated device, such as the miniextruder supplied by Avanti Polar Lipids, through polycarbonate filters of pore sizes 0.6, 0.4, and 0.2 μm. This procedure was used in some studies on the effect of extrusion on particle size and polydispersity of the final liposomal preparation. The proliposomes are extrudable up to a pore size of 0.2 μm. Higher pressures during the extrusion through a 0.1-μm pore size filter or ultrasonication result in disorganization of proliposomes (tightly packed bilayer sheets) and their transformation into a viscous dispersion of phospholipids. Proliposomes represent transient metastable structures and a high pressure probably induces dehydration of bilayers and metamorphosis back into the amorphous state.

We have found that extrusion of proliposomes influences both the final size and polydispersity of the resultant liposomes. The smaller the filter pores used, the smaller the size and the lower the polydispersity of liposomes obtained. The results obtained by electron microscopy (Fig. 6) and light-scattering measurement (Fig. 8) show a good accordance.

Sterility Testing

The cell is sterilized with 60% ethanol at 60° for 15 min before use. *Staphylococcus aureus* (A-positive strain 722 obtained from the Food Research Institute, University of Wicsonsin, Madison, WI) is maintained in glycerol broth at −20° and 24-h-old cultures grown on blood agar plates at 37° are used as a model for experimental contamination of the cell (10^6 CFU) before sterilization. The sterility of the product is checked by culture on blood agar plates (37°, 96 h). No bacterial contamination was detectable after 96 h of culture on blood agar in any of the liposomal preparations under study.

Concluding Remarks

We present a description of a stirred thermostatted cell and its link-up with a liquid delivery system for the rapid production of multilamellar liposomes by the proliposome–liposome method, which is based on the conversion of the initial proliposome preparation into a liposome dispersion by dilution under strictly controlled conditions. The design of the cell allows easy assembly and link-up with FPLC or other delivery systems that facilitate full control of the process, giving highly reproducible results. Most components used in the construction can be found in any laboratory or

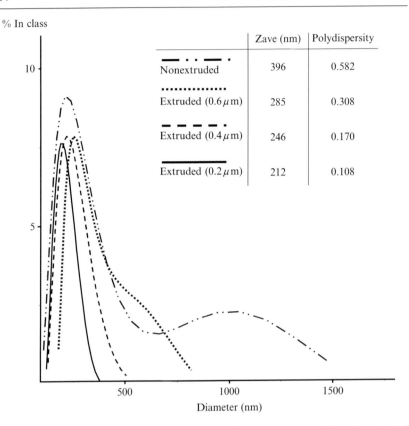

	Zave (nm)	Polydispersity
Nonextruded	396	0.582
Extruded (0.6 μm)	285	0.308
Extruded (0.4 μm)	246	0.170
Extruded (0.2 μm)	212	0.108

FIG. 8. Size distribution of liposomes prepared from nonextruded and extruded proliposomes.

are listed in catalogs of major suppliers of laboratory utensils. Larger quantities of proliposomes (above 1 ml in total volume, 300–1000 mg of phospholipids) can be easily prepared directly in the cell by mixing an ethanol solution of phospholipids with the water phase at the desired flow rate and temperature. The liposomes are produced according to good manufacturing practice (GMP) and are sterile. The cell has been designed for laboratory-scale preparation of liposomes (300–1000 mg of phospholipid per run) by a procedure taking less than 90 min. The method can be readily scaled up and linked with secondary processing methods, such as extrusion through polycarbonate filters for the preparation of oligolamellar and unilamellar liposomes.

The cell is a prospective tool for modeling of processes of industrial-scale preparation of liposomes.

Vesiculation of lipid bilayers and sealing of the membrane are critical steps in the formation of liposomes. Unstable transient structures appear during the formation of liposomes by a process that is difficult to investigate. A unifying model, based on a bilayered phospholipid fragment, has been proposed by Lasic.[17] No data describing the mechanism of transformation of proliposomes (dispersed floccules of lamellar phase) into liposomes have been found in the available literature. Also unclear are the existence and possible role of bilayered phospholipid fragments in that process. We assume that a relatively high lipid concentration and the large surface area of the lamellar phase in the proliposome mixture are responsible for high entrapment efficiencies for various drugs.

Variations of some parameters influencing the structure of proliposomes (e.g., phospholipid composition, addition of small amount of detergent, temperature, pH and ionic strenth, and addition of glycerin or saccharides) and the process of vesiculation (decreasing of ethanol concentration by, e.g., dilution, dialysis, or evaporation at lower presure) can release a hidden potential of versatility of this method with respect to size, morphology, and entrapment efficiency of final liposomal preparations.

The method has been successfully applied in our laboratory for the encapsulation of various drugs. Prospective candidates for entrapment into liposomes prepared by this technique are particularly lipophilic derivatives (e.g., modified by long acyl chain, cholesterol, or phospholipid) of muramyl dipeptide with immunostimulating activities.[18] This method is also applicable to large-scale preparation of cationic liposomes and plasmid DNA–liposome complexes for DNA vaccine preparation.

Acknowledgments

This work was supported by the Ministry of Agriculture of the Czech Republic (MZE-M03-99-01). We thank M.V. Dr. Bedřich Šmid, Dr. Sc., for preparation of the electron micrographs of liposomes, Andrea Tománková and Irena Trnečková for preparation of pictures, and Mr. Jindřich Prokeš for manufacturing some parts of the stirred cell.

[17] D. D. Lasic, *Biochem. J.* **256**, 1 (1988).

[18] J. Turánek, D. Záluská, M. Hoffer, A. Vacek, M. Ledvina, and J. Ježek, *Int. J. Immunopharmacol.* **19**, 611 (1997).

Section II

Physicochemical Characterization of Liposomes

[10]   Phase Behavior of Liposomes

By Paavo Kinnunen, Juha-Matti Alakoskela,
and Peter Laggner

Introduction

With respect to their chemical structure the diversity of lipids is impressive, with more than 1000 different species being present in an average eukaryotic cell. Although understanding of the functional significance of the various lipids is still in its infancy, it is important to recognize that the patterns of lipids found in different cell types and further in subcellular organelles exhibit striking specificities. The maintenance of this diversity requires a wealth of metabolic energy as well as extensive enzymatic machineries encoded by the genome. Accordingly, this expenditure must yield the cells significant advantages and play central roles in vital processes. Examples of the latter are illustrated by the involvement of lipids such as phosphatidylinositols and ceramide in cellular signaling cascades.

Biochemistry of lipids mainly describes their metabolic processing, with cell biology emphasizing the phenomenology of the dynamic distribution of lipids in cells. Understanding of the molecular basis of lipid functions requires biophysical approach. To this end, a key feature is that lipids represent the liquid crystalline state of matter, characteristically exhibiting a number of different phases, connected by phase transitions. These phase changes are triggered by factors such as temperature, ions, electric fields, and pH, to name a few, and cause dramatic alterations in the physical properties and organization, and therefore also the functions, of any multicomponent system involving lipids.[1] Although ordering on different time and length scales is inherent to a many-body system such as is represented by the various cellular membranes,[2] it is still the physical properties and phase behavior of the individual lipids that determine the final outcome. Moreover, close to critical points these systems become inherently unstable to minor impurities and even low concentrations of membrane-perturbing agents, such as drugs and metabolites, can have drastic impact on membrane dynamics on different levels.

---

[1] P. K. J. Kinnunen and P. Laggner, eds., *Chem. Phys. Lipids* **57,** 109 (1991).

[2] O. G. Mouritsen and P. K. J. Kinnunen, *in* "Biological Membranes: A Molecular Pespective from Computation and Experiment" (K. M. Merz and B. Roux, eds.), p. 463. Birkhäuser, Boston, 1996.

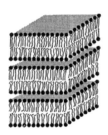

FIG. 1. Examples of three common lyotropic structures of lipids. *Left:* Lamellar (L_α). *Center:* Hexagonal (H_{II}). *Right:* Cubic (*Pn3m*). From Ref. 5.

Key to the description of lipid phase behavior is composing the phase diagram. Routinely, this involves causing phase changes by temperature, while varying the chemical composition of the system. The methods required are differential scanning calorimetry (DSC) yielding thermodynamic data on the system, in combination with methods providing structural information, such as diffraction techniques (neutrons and X-rays) and nuclear magnetic resonance (NMR). More detailed views can be obtained when these approaches are complemented by computer simulation, Langmuir balance studies, fluorescence, and vibrational spectroscopies, for example. It is, however, important to recognize that this approach, although providing fundamental description of the system, is manageable only when dealing with a limited number of components. Extrapolation on the basis of these data to real, many-component systems is still far away.

A fundamental division of the different phases formed by lipids is provided by their three-dimensional (3-D) nature. Accordingly, the phases are described as lamellar, hexagonal, and cubic (Fig. 1). A detailed description of these phases is beyond the scope of this chapter and the interested reader is referred to available reviews and monographs.[3-7] Finally, it should be mentioned that rational description of these phases and phase

[3] D. Marsh, *Chem. Phys. Lipids* **57,** 109 (1991).

[4] M. W. Tate, E. F. Eikenberry, D. C. Turner, E. Shyamsunder, and S. M. Gruner, *Chem. Phys. Lipids* **57,** 147 (1991).

[5] J. M. Seddon and R. H. Templer, *in* "Handbook of Biological Physics" (R. Lipowsky and E. Sackmann, eds.), vol. 1A, p. 97. Elsevier, Amsterdam, 1995.

[6] G. Lindblom and L. Rilfors, *Biochem. Biophys. Acta* **988,** 221 (1989).

[7] J. M. Seddon and G. Cevc, *in* "Phospholipids Handbook" (G. Cevc, ed.), p. 403. Marcel Dekker, New York, 1993.

transitions is not only needed to develop an understanding of the functions of biomembranes, but is also crucial for the development of biotechnological applications of lipids, such as the use of liposomes for targeted drug delivery. Other potential areas are the use of cationic lipids as transfection vectors, to convey desired DNA into cells,[8] and transdermal administration of therapeutic agents.

Preparation of Liposomes

Sample preparation is the most critical part in the investigation of liposomes. In comparison with spectroscopic or calorimetric techniques, which are most sensitive to short-range molecular packing and less affected by long-range morphological variations or defects, it is one of the strengths of diffraction methods, and in particular of SWAX (simultaneous small- and wide-angle X-ray diffraction[9,10], to be sensitive to subtle morphological variations. These may, besides compositional factors, be affected by the mode of preparation. SWAX data on liposomes, with their detailed line widths and shapes, are reliable only if one and the same procedure is followed carefully throughout an entire series. Of the many variables that might play a role, the most frequent sources of problems are briefly discussed in the following.

Lipid Purity

Lipid stock solutions are made in a proper organic solvent such as chloroform or chloroform–methanol and are stored protected from light at $-20°$. The purity of the lipids should be checked periodically for possible decomposition by thin-layer chromatographic (TLC) analysis. Apart from lipidic impurities, which can be detected by TLC and removed by suitable methods (preparative TLC or high-pressure liquid chromatography), low molecular weight inorganic contaminants can cause serious problems as they may affect the solubility of the lipids. Such contaminants are, for instance, bromide or iodide carried over from the catalysts used in lipid synthesis. Even if present only to a few percent by weight, they represent a major impurity on a mole fraction basis. Finally, in addition to purity, it is advisable to monitor the concentrations of the stock solutions so as to correct for the unavoidable solvent evaporation. Most convenient

---

[8] P. L. Felgner, T. R. Gadek, M. Holm, R. Roman, H. W. Chan, M. Wenz, J. P. Northrop, G. M. Ringold, and M. Danielsen, *Proc. Natl. Acad. Sci. USA* **84,** 7413 (1987).

[9] P. Laggner and K. Lohner, in "Lipid Bilayers: Structure and Interactions" (J. Katsaras and T. Gutberlet, eds.), p. 233. Springer-Verlag, Berlin, 2000.

[10] P. Laggner and H. Mio, *Nucl. Instrum. Methods Phys. Res. A* **232,** 86 (1992).

for this purpose is gravimetric analysis, using a high-precision laboratory balance.

Lipid Dispersions

To make liposomes the lipid mixtures or lipids and other components soluble in organic solvents are mixed in the said solvent or solvent mixture to obtain the desired composition as an optically clear solution. The glassware used should be carefully cleaned by rinsing with ethanol, for instance, to ensure the absence of any surface-active materials, such as detergents. Mixing of the lipids is followed by removal of the solvents by evaporation, facilitated by gentle blowing of nitrogen (oxygen free to avoid lipid peroxidation) over the sample. For unsaturated lipids protection from exposure to light is recommended. Subsequently, the dry lipid film is kept under reduced pressure for at least 2 h (preferably overnight), so as to eliminate the presence of residual solvents.

The drying procedure leads necessarily to a demixing. First, because the solvent components have different vapor pressures, and evaporate at different rates until an azeotropic mixture is reached, it is likely that the lipid components have different solubilities in the remaining azeotrope, which in the case of chloroform–methanol, for example, is highly enriched in the latter, which is the more polar solvent. Second, even if the lipid components are soluble in this azeotrope, they will have different solubility limits as their concentration increases. In the case of chloroform–methanol, therefore, the more apolar component (e.g., cholesterol) will precipitate first. In the case of a pure solvent, again, the component for which this is the better solvent will stay in solution, while the other already starts to precipitate. A facilitated precipitation of a lipid mixture or complex can happen, but inevitably the dry residue will be initially demixed. A remixing of the components can occur only via diffusion, and therefore the incubation of the dry residue at elevated temperature, above the chain-melting transition of the lipids, may help. The next step is to disperse the lipids in an appropriate volume of water or buffer, and to homogenize the sample, again at a temperature at which the membrane or its highest melting point lipid component is in the liquid crystalline, disordered state. Knowing this temperature thus requires preliminary DSC data to be recorded before the actual analysis. In these preliminary experiments the effect of heating rate can be verified.

Dispersion of the sample can be aided by vortexing, or if the sample is too concentrated and viscous, by centrifuging many times back and forth through a narrow constriction in a thick-walled glass tube. Before the DSC and X-ray measurements are started, the samples must be equilibrated at the desired temperature, or at the starting temperature of a series. The incubation of the samples at reduced temperatures ($+4°$) before the heating scan is important, as the relaxation of the assembly to the lowest free energy state may require exceedingly long times, in some cases weeks and months. An example of a lipid exhibiting such behavior is dimyristoylphosphoglycerol.[11,12] In this connection it is worth mentioning that the possible biological significance of these metastable phases is completely unknown. In any case, it must be checked by repeated measurements under different conditions that equilibrium or a stable state has been reached.

Unilamellar Lipid Vesicles

Liposome phase behavior commonly refers to micron-sized, multilamellar dispersions. For studies on single bilayer membranes and their interactions, different standard methods for the preparation of vesicles, that is, quasi-hollow spherical structures confined by a continuous bilayer, have been developed and are widely in use. For large unilamellar vesicles (LUVs) the method of pressure extrusion through defined pore-size filters [Lipoprep (Harvard Bioscience, Holliston, MA), Nuclepore filters (Whatman Nuclepore, Scarborough, MA)] results in vesicles approximately 100 nm in diameter,[13,14] while ultrasonication[15,16] yields small unilamellar vesicles (SUVs) with an average diameter of 25 nm. The sizes and stability of both types of vesicles vary depending on composition, temperature, and time. SUVs are generally metastable and tend to fuse, so that careful checks on their viability by, for example, light scattering or other particle sizing methods are advisable. As far as their single- or oligoshell nature is concerned, X-ray small-angle scattering can provide relevant information by the fact that unilamellar lipid vesicles lead to smooth and weak scattering curves, whereas oligo- or multilayer liposomes show progressively sharp Bragg peaks (Fig. 2).

[11] I. S. Salonen, K. K. Eklund, J. A. Virtanen, and P. K. J. Kinnunen, *Biochim. Biophys. Acta* **982**, 205 (1989).

[12] H. W. Meyer, W. Richter, W. Rettig, and M. Stumpf, *Colloids Surf. A* **183–185,** 495 (2001).

[13] L. D. Mayer, M. J. Hope, and P. R. Cullis, *Biochim. Biophys. Acta* **858**, 161 (1986).

[14] R. C. MacDonald, R. I. McDonald, B. M. Menco, K. Takeshita, N. K. Subbarao, and L. R. Hu, *Biochim. Biophys. Acta* **1061**, 297 (1991).

[15] J. T. Mason and C. Huang, *Ann. N. Y. Acad. Sci.* **308**, 29 (1978).

[16] F. Szoka and D. Papahadjopoulos, *in* "Liposomes: From Physical Structure to Therapeutic Applications" (C. G. Knight, ed.), p. 51. Elsevier, North-Holland Biomedical Press, Amsterdam, 1981.

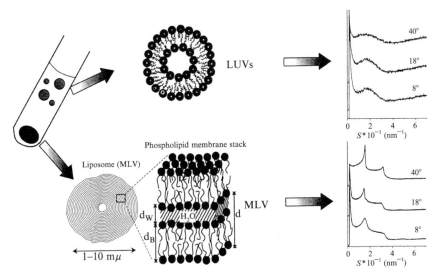

FIG. 2. Typical x-ray small-angle data as obtained from unilamellar vesicles and multilamellar liposomes, respectively, obtained in one preparation of a lipid–peptide (DMPC–δ-lysin) preparation. Continuous scattering curves are obtained from the vesicles, whereas sharp Bragg peaks originate from multilamellar liposomes (MLVs). From Ref. 9.

Differential Scanning Calorimetry

The principle of this method is to measure the amount of energy needed to elevate the temperature of a sample at a given, steady rate. The sensitivity of the current commercially available instruments is excellent and allows the use of lipid concentrations as low as $20\,\mu M$. Accordingly, samples with excess water can be used, which is important when lyotropic, that is, concentration-dependent, phase changes are not being investigated. These instruments also offer a wide range of heating rates. The selection of the latter is important as the kinetics of phase changes can be slow. Too-fast heating of such compositions takes them into metastable states, not representing thermodynamic equilibrium. Accordingly, a slow rate should be selected to monitor the development of the transitions as close to equilibrium as possible.

The actual measurement with a modern commercial instrument and its dedicated software is straightforward. The cleanliness of measurement cells must be emphasized, however. The cells should be rinsed with at least 0.5 Liters of water, followed by ethanol between the measurements. Subsequently, it is necessary simply to fill the cell with the lipid sample in buffer and the reference measurement cell with plain buffer. Importantly, before loading the samples the DSC cuvettes must be cooled to the desired starting temperature and care must be taken so as to prevent warming of the lipid dispersions during their transfer into the cuvettes. Subsequently, the measurement parameters in the program controlling the instrument are set. These parameters include the prescan incubation time, scan rate, and incubation time between upscan (heating) and downscan (cooling) if both are to be measured. A short (10-min) prescan incubation is sufficient to ensure that the sample temperature has reached equilibrium with the surroundings. It is vital that the thermal history for each sample is identical unless, of course, metastability is the subject of the study. The presence of metastable states can also be revealed by hysteresis or the different temperatures for the transition in upscan and downscan.[17] Because of this, the incubation time between upscan and downscan could also have significance with respect to the amount of metastable states remaining in the sample, but usually a short incubation of 0.5 h should be sufficient.

After the samples have been applied to the instrument, the sample compartment is adjusted to a constant pressure, this pressure being usually somewhat higher than the atmospheric pressure in order to avoid bubbling during heating (0.2 MPa was used for Figs. 3 and 4). Some commercial instruments with pressure perturbation accessories allow the measurement of volumetric changes at transitions, but thanks to their slightly extended pressure range they also allow the study of effects of applied pressure on the phase transition temperature.[18]

Dedicated software, available from instrument manufacturers, automatically converts the energy–time and temperature change–time scales into raw data as heat capacity versus temperature. After the raw data have been obtained, one should either first subtract the baseline or divide by the lipid concentration in order to obtain molar quantities instead of bulk quantities (Fig. 3A). The order in which these operations are done has no significance. Usually for lipids as well as for most other substances the molar heat capacity has a weak dependency on temperature outside the temperature ranges of the transitions. Therefore the baseline should be almost horizontal unless (most likely) experimental artifacts impose a time- or temperature-dependent tendency on the data (Fig. 3A). Most often the baseline can be set to zero by simple subtraction of a straight line (Fig. 3). More complicated baseline curves can, of course, be used, but especially in composing lipid phase diagrams it is best to consider both the

[17] B. Tenchov, *Chem. Phys. Lipids* **57,** 165 (1991).
[18] H. Heerklotz and J. Seelig, *Biophys. J.* **82,** 1445 (2002).

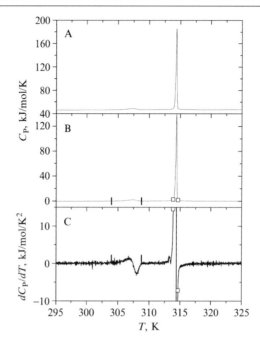

FIG. 3. (A) Raw data from a DSC measurement of dipalmitoylphosphatidylcholine (DPPC) multilamellar vesicles (MLVs). Only concentration normalization has been done, that is, each heat capacity value has been divided by a DPPC concentration of 1.7 mM. The scan rate was 20 K/h. (B) Straight line has been subtracted to adjust the baseline. The vertical bars and open squares indicate the start and end points for pretransition and main transition, respectively, as found by Origin 4.2 for the DSC peak-finding algorithm. (C) The numerical derivative of the data in (A). The vertical bars and open squares indicate the start and end points for the transitions as found by the peak-finding algorithm.

effects this could have on the shape of the transition peak and the nature of the process leading to these complicated baseline curves.

When beginning to analyze the DSC curves (Fig. 3A) it is possible to extract the temperatures for the start point of the transition (where the curve first deviates from the baseline), the end point of the transition (where the curve returns to the baseline), and the temperature at which the heat capacity has a maximum. The last of these values is usually referred to as the transition temperature in experimental publications, including this chapter. Alternatively, the term transition temperature is sometimes assigned to the enthalpy midpoint of the transition, that is, the temperature that divides the peak into two parts having equal areas. Finally, when there is considerable hysteresis, the term transition

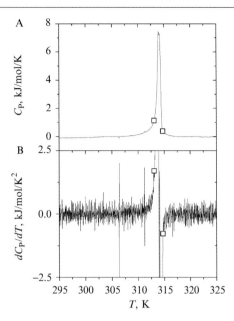

Fig. 4. (A) Baseline-subtracted data from a DSC measurement of DPPC extruded large unilamellar vesicles (LUVs). The open squares indicate the start and end points found by the peak-finding algorithm. The scan rate was $20\,K/h$. (B) The numerical derivative of the raw data for LUVs. Open squares as in (A).

temperature in theoretical papers is sometimes assigned to the intersection temperature of the heating and cooling scans.[19] The different definitions tend to produce slightly different values, as for phospholipids the peaks are usually asymmetric and there often is some hysteresis. Yet, once the term transition temperature has been defined, the value can be determined unambiguously. In contrast, slightly different results are obtained easily when trying to determine the exact start and end points of the transition. Visual inspection is the most common method, and a good one, if the interpretations of the experimenter remain uninfluenced by hopes and opinions. Yet, as the peaks curve smoothly down to the baseline, it is easy to obtain as much as $0.5°$ differences depending on the interpretation, or even a few degrees for smooth transitions such as the phospholipid pretransition. However, considerable aid for visual inspection is provided by the numerical derivative of the heat capacity curve (Fig. 3C).

[19] I. P. Sugar, *J. Phys. Chem.* **93,** 5216 (1989).

In addition to visual inspection it is possible to use peak-finding algorithms included in DSC programs or criteria such as the extent of deviation from the baseline compared with the noise level. The great advantage of these methods is that unlike visual inspection they are user independent, so they find the baseline deviation points by applying identical rules for each measured trace. Unfortunately, however, they often tend to produce results that clearly do not match the temperatures at which the curve begins to deviate from baseline and returns to baseline (Figs. 3 and 4). Thus, these results probably do not coincide with the physical reality of phase coexistence temperature range.

Both the enthalpy and entropy associated with a phase transition can be calculated numerically on the basis of the experimental DSC excess heat capacity-versus-temperature curves. As the pressure is kept constant, it corresponds to the heat capacity at constant pressure, C_P, due to lipids, defined as

$$C_P = \left(\frac{\partial H}{\partial T}\right)_P \tag{1}$$

The enthalpy H associated with the transition is obtained by integrating the experimental C_P-versus-T curve over the temperature range of the transition, that is, calculating the area of the peak. From the definition of enthalpy, $H = E + PV$ (internal energy plus pressure multiplied by volume), and the fundamental equation of thermodynamics, $dE = T \times dS - P \times dV$, one obtains the differential of enthalpy, $dH = T \times dS + V \times dP$. At constant pressure $dP = 0$, which leads to $dS = T^{-1} \times dH$. As seen from Eq. (1), $dH = C_P \times dT$, and therefore $dS = T^{-1} C_P dT$. By integrating both sides one obtains the value for the change of entropy, ΔS, associated with changing the system from entropy S_1 and temperature T_1 to entropy S_2 and temperature T_2:

$$\Delta S = S_2 - S_1 = \int_{S_1}^{S_2} dS = \int_{T_1}^{T_2} \frac{1}{T} C_P dT \tag{2}$$

By selecting the temperatures to correspond to the start and end points of the transition, one obtains the change in entropy associated with the phase transition. When the temperature interval over the peak is narrow compared with the actual temperatures, that is, when $T_1 \approx T_2$, one can make the approximation of constant transition temperature T_t over all the interval and obtain

$$\Delta S = \frac{\Delta H}{T_t} \tag{3}$$

For phospholipids, even as mixtures, transitions do not extend over too wide a temperature range (about a few degrees). Considering the magnitude of transition temperatures ($\sim$300–350 K), the error using approximation (3) is usually negligible, and therefore the easiest way to calculate the change in entropy is to divide the change in enthalpy by the transition temperature.

X-Ray Diffraction: Methods and Examples

With modern instrumentation, x-ray diffraction methods have reached a speed and precision well suited for mapping the phase behavior of single- and multicomponent lipid systems. Single measurements are performed in minutes, and with suitable automation tools several dozens of samples can be measured in the course of one working day. The time-limiting factor is the preparation and equilibration of the samples and not the measurement itself. High-throughput screening, or time-resolved observation on annealing of liposome formulations, has become an easy task even with conventional laboratory x-ray sources.[20] Figure 5[21] shows a typical example of a temperature scan, by small- and wide-angle x-ray diffraction (SWAX), of a multilamellar lipid vesicle (MLV) preparation.

The type of structural information that can be obtained from such measurements is summarized in Table I. In general, the decision about the overall phase structure, whether it is lamellar, hexagonal, or cubic, can be made by visual inspection, or by employing simple autoindexing software routines. The following simple relations for the positions of the Bragg peaks give the fingerprint for the most frequently observed lipid mesophases:

Lamellar: 1, 2, 3..., ..., n
Hexagonal: 1, $\sqrt{3}$, 2, $\sqrt{7}$, 3 ...
Cubic: 1, $\sqrt{2}$, $\sqrt{3}$, 2, $\sqrt{8}$, ... (depending on space group)

Equally, the chain-packing mode, crystalline, gel, or liquid crystalline, is seen easily from the wide-angle patterns: the orthorhombic chain-packing lattice shows a prominent peak at 4.18 $\overset{\circ}{A}^{-1}$ and a shoulder around 4.08 $\overset{\circ}{A}^{-1}$, the hexagonal packing shows a peak around 4.11 $\overset{\circ}{A}^{-1}$, whereas the liquid paraffin chains show a broad peak around 4.5 $\overset{\circ}{A}^{-1}$.

[20] With the high x-ray flux of synchrotron x-ray sources, the measuring times have come down to fractions of milliseconds, allowing real-time cinematographic studies of the nonequilibrium dynamics and pathways of structural transitions; for reviews, see P. Laggner and M. Kriechbaum, *Chem. Phys. Lipids* **57**, 121 (1991); M. Caffrey and A. Cheng, *Curr. Opin. Struct. Biol.* **5**, 548 (1995); and P. Laggner, *in* "Spectral Properties of Lipids" (R. J. Hamilton and J. Cast, eds.), p. 327. Academic Press, Sheffield, 1999.

[21] K. Pressl, K. Jørgensen, and P. Laggner, *Biochim. Biophys. Acta* **1325**, 1 (1997).

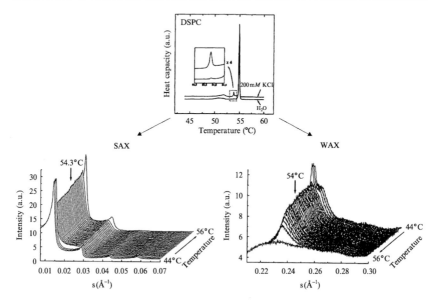

Fig. 5. High-resolution temperature scanning of 1,2-distearoyl-*sn*-glycero-3-phosphocholine (DSPC) liposomes in the pre- and main transition region. *Top*: DSC scans. *Bottom left*: Small-angle x-ray data. *Bottom right*: X-ray wide-angle data taken simultaneously in the SWAX camera (see Fig. 6). The intensity shifts around 54° show the submain transition.[21]

For the purpose of evaluating phase diagrams, automatic scanning procedures, in which series of diffraction patterns are measured as functions of chemical (concentration, cosolute, etc.) or physical (temperature, pressure) variables, are convenient. They not only serve to find the boundaries for the phase diagram, but also characterize the phase structures. Linking this information to calorimetric data provides a solid basis for the interpretation of the thermodynamic state functions, enthalpy and entropy. It is important to note that the simultaneous observation of both the chain packing from WAX and the long-range structure from SAX is a central technical issue to such an approach.

Detailed analysis of individual diffraction patterns in terms of bilayer structure, dynamic and static disorder, domain formation, and interactions with membrane-active agents requires a higher level of sophistication and often must resort to the combination with information from other sources (e.g., x-ray diffraction from oriented samples, neutron scattering, and imaging techniques). Through the ease and speed of measurement brought about by modern SWAX instruments it is advisable to include general chemical and physical intelligence to the experimental

TABLE I
INFORMATION CONTENT OF X-RAY DIFFRACTION ON LIPOSOMES

Information	Type of experiment/analysis
Phase structure (lamellar, hexagonal, cubic, etc.) and lattice parameters, such as lamellar repeat distance	Small-angle powder diffraction (SAX), indexing of Bragg reflections
Mode of chain packing (crystalline, tilted, gel, or liquid-crystalline), unit cell area, etc.	Wide-angle powder diffraction (WAX), indexing of Bragg reflections
Bilayer dimensions (thickness, area per molecule), solvation and mechanical properties (bending elasticity modulus, compressibility, undulation)	Quantitative, model-dependent analysis of single SAX patterns (Fourier analysis, deconvolution[a]), e.g., by the MCG method[b]
Lattice disorder, defects, phase transitions/separations	Line-shape analysis of SAX patterns, Porod analysis
Bilayer dimensions in single-shelled vesicles, electron density profile	Fourier transformation of continuous SAX curves, deconvolution

Abbreviations: MCG, modified Caille-Gauss.

[a] P. Laggner, *in* "Topics in Current Chemistry" (E. Mandelkow, ed.), vol. 145, p. 173. Springer-Verlag, Heidelberg, Germany, 1988.

[b] G. Pabst, M. Rappolt, H. Amenitsch, and P. Laggner, *Phys. Rev. E* **62,** 4000 (2000).

strategy: rather than squeezing an individual diffraction pattern to the extreme by model-dependent analysis, it is often more promising to obtain useful information by mapping out the diffraction properties under a range of different conditions. Such information has often higher value concerning the biophysical question behind the study than a set of abstract structural parameters.

Instrumentation

The main elements of an x-ray camera are the following:
Source: X-ray generator, or synchrotron beam
Collimator, to provide a well-defined, narrow beam
Sample stage
Detector(s), data acquisition system
As laboratory x-ray sources, both sealed-tube and rotating-anode generators are suitable and widely in use. For collimation, two principal geometries can be chosen: line collimation or point collimation. For randomly oriented systems such as liposomes, in which the scattering is isotropic

about the direction of the primary beam, the choice of a line collimator is of advantage over point collimation, for reasons of intensity, resolution to small angles, and costs (linear position sensitive detectors are much less expensive than electronic area detectors). High-precision systems, such as the Kratky collimator, can provide a small-angle resolution, that is, the angle closest to the primary beam direction, of about 1 mrad, corresponding to d values of 150 nm, well suited for liposome studies. Only with oriented specimens, such as oriented lipid films at a solid surface, may point collimation be needed, although with lamellar systems line collimation can also be employed.

The sample holder consists typically of a 1-mm-diameter glass or quartz capillary [wall thickness, 0.01 mm (e.g., Mark capillaries; Hilgenberg, Malsfeld, Germany)] for liquid samples, or a vacuum-tight Teflon/stainless steel holder for pastes and powders with x-ray windows, which can be either polymers (Mylar or Kapton foils; note, however, that these materials have a certain vapor permeability) or mica. The sample volumes necessary for a sealed capillary are about 10–20 μl, and for automatic pump feeding about 200 μl. The capillary holder should be fitted to a vacuum-tight feed-through system that allows automatic, pump-driven sample changes for serial measurements. A precise and reliable temperature control unit with PC programmability is a central requirement for large-scale phase diagram mapping. When the sample contains crystalline macrodomains, as are often found with cubic phases, it is necessary to rotate the capillary in a specially designed cell (e.g., Spin-Cap; Hecus M. Braun, Graz, Austria) during measurement.

The x-ray detectors most widely used in combination with the previously-described optics are linear, position-sensitive single-photon counters, with a nominal pixel resolution of about 50 μm. The maximal local (peak) and global (flat illumination) count rates are about 100 and 50,000 s^{-1}, respectively, which is commensurate with the typical intensities observed from liposome preparations with anode powers of less than 6 kVA. Higher intensities lead to coincidence losses in the gas detectors, which may seriously distort the diffraction patterns. In this context it must be noted that the quality of primary beam-stop, which shields off the damaging high intensity of the direct x-ray beam from the detector at the smallest possible angles, is of particular relevance. In modern camera systems, this is an x-ray absorber located closely in front of the detector window, within the vacuum beam path of the camera. The absorber can be adjusted with micrometer precision in the primary beam position, and facilitates the measurement of the primary beam position and relative intensity simultaneously with the small-angle diffraction.

For simultaneous small- and wide-angle measurements, two detectors are coupled to the central data acquisition system (SWAX camera; Hecus M. Braun). In that configuration, the detectors are time multiplexed, with a free choice as to individual acquisition times (down to millisecond time frames) for the small- and wide-angle detectors. This allows for obtaining optimal counting statistics in both SAX and WAX patterns, which may differ in their intensities. A camera widely used in this field is shown in Fig. 6.

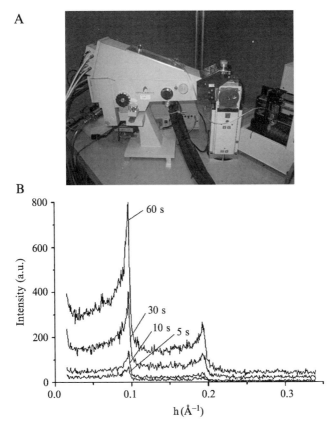

FIG. 6. (A) Small- and wide-angle x-ray camera (SWAX; Hecus M. Braun, Graz, Austria) for automatic, serial measurements with flow cell and autosampler. (B) Typical results from 20% liposome dispersions (1-palmitoyl-2-oleoyl-sn-glycero-3-phosphocholine, POPC) taken at various exposure times. Generator power, 2 kVA. At 10 s, the intensities and peak positions are already well resolved.

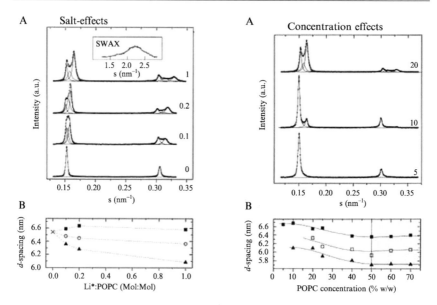

FIG. 7. Effects of alkali salt (LiCl) on the L_α phase of palmitoyl-oleoyl phosphatidylcholine (POPC) liposomes. *Left*: Variation of salt:lipid ratio. *Right*: Effects of POPC concentration. From Ref. 22.

Salt Effects

Salts, in particular alkali and earth alkali halides, have pronounced effects on the solvation properties of lipid bilayers, and this possibility must be taken into account when planning an experimental protocol. Figure 7[22] shows SAX evidence of the coexistence of different L_α phases in a typical liposome preparation in distilled water and in the presence of increasing concentrations of LiCl. Although this may not be relevant in a rough, cursory examination of the phase state, it is obviously crucial if the lipid–salt interaction is the objective of the investigation. In the latter case the method of preparing the appropriate lipid–salt–water concentration systems is critical. In practice, there are two alternatives: either dispersing the dry lipids in the desired aqueous salt solution, or adding concentrated salt solution to a lipid dispersion. Although both methods should lead to the same equilibrium state, this is in practice not always the case.[23] This indicates that insufficient time has been allowed for equilibration.

[22] M. Rappolt, K. Pressl, G. Pabst, and P. Laggner, *Biochim. Biophys. Acta* **1372**, 389 (1998).
[23] The matter is further complicated by the fact that different metastable states may be obtained depending on whether the dry lipids are crystalline (from commercial stock) or prepared by solvent evaporation.

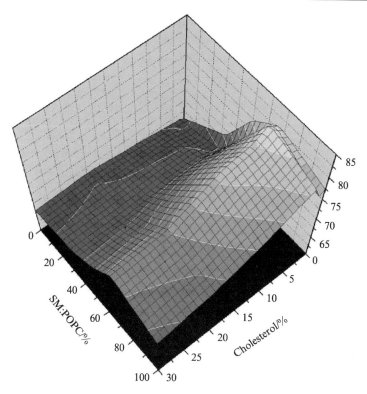

Fig. 8. Typical result from an x-ray phase-mapping series. System: POPC–sphingomyelin–cholesterol. The *d*-spacings of the first-order Bragg peak show a profile with a maximum around molar ratios of 3:1:0.1.

Equilibrium may in certain cases not be reached at all in reasonable time spans, and the methods lead to different metastable states. This obviously calls for great care concerning reproducibility of protocols.

Phase Diagram Mapping

In Fig. 8, a representative example of the potential of x-ray diffraction in phase diagram mapping of a pseudo–three-component system (the fourth component, water, is in large excess) is shown. In this example, the Bragg spacings show a maximum at a certain phosphatidylcholine:sphingomyelin:cholesterol ratio, which is likely to be caused by a maximum in hydration and/or bilayer undulation. It should be noted that this type of information about variations in the phase structure within one given phase region (e.g., gel or liquid crystalline) can be obtained only from x-ray diffraction. It is

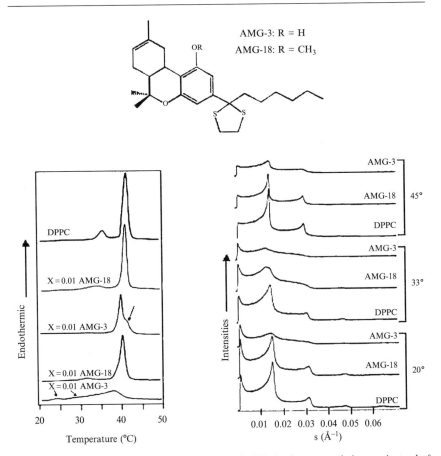

FIG. 9. Effects of cannabinoid model drugs (*top*), differing by one methyl group instead of a hydrogen atom, on the phase transition behavior and structure of DPPC liposomes. *Bottom left*: DSC scans. *Bottom right*: SAX patterns at various temperatures corresponding to the $L_{\beta'}$, the $P_{\beta'}$, and L_α phases, respectively. From Ref. 24.

also clear that without a computer-controlled autosampler and measurement guidance, such a comprehensive survey would be laborious. With automated techniques, however, this can now be accomplished as a routine task.

Another example for an important field of application is shown in Fig. 9,[24] where the membrane activity of two different, but chemically closely related, cannabinoid compounds is examined. One of them,

[24] T. Mavromoustakos, D. Papahatjis, and P. Laggner, *Biochim. Biophys. Acta* **1512**, 183 (2001).

AMG-3, is pharmacologically active, and the other, AMG-18, is not. Here again, the combination of calorimetric and x-ray techniques leads to a detailed picture of the thermodynamic and structural events in the interaction. Accordingly, the pharmacologically active AMG-3 efficiently fluidizes the lipids in the $L_{\beta'}$ gel phase and perturbs the regular multilayer lattice. In the liquid crystalline L_α phase, AMG-3 is also much more effective in disrupting the multilayer regularity, whereas AMG-18 has only minor effects. These dose-dependent, different thermodynamic and structural effects by the two cannabinoids suggest that these may be related to their biological activity.

Concluding Remarks

The present review has given a brief summary of the most general experimental approaches, as practicable in state-of-the-art laboratories, to the mesomorphic phase behavior of liposomes. The techniques of calorimetry and x-ray diffraction have reached a state of simplicity in operation, making them routine tools in any R&D or QC application in this rapidly advancing field of technology. For the sake of general usefulness and space, important specialities, such as time-resolved x-ray diffraction for studies on nonequilibrium states, surface diffraction on solid-supported model membranes, or relaxation calorimetry have been left out. This also reflects the authors' belief that in the present state of knowledge, it is of paramount importance to focus on a broad landscape mapping of the structural and thermodynamic properties of multicomponent lipid systems. In this pursuit, it is most important to combine biochemical and biophysical data with modern high-throughput techniques, for which the diffraction methods described here are already well suited; the calorimetric methods need to be speeded up. Yet, even partial phase diagrams obtained for model membranes can provide valuable insight into the changes in the organization and properties of biomembranes, as exemplified by comparison of the role of ceramide in apoptosis with the biophysical properties of this lipid.[25]

[25] J. M. Holopainen, J. Lemmich, F. Richter, O. G. Mouritsen, G. Rapp, and P. K. J. Kinnunen, *Biophys. J.* **78,** 2459 (2000).

[11] Electrophoretic Characterization of Liposomes

By JOEL A. COHEN

Introduction

Liposomes are colloidal particles composed of lipids. The physical properties of liposomes can vary widely, depending on their lipid composition and method of preparation. Perhaps the most fundamental physical parameters characterizing a liposome are its size and its intrinsic electrical charge or surface charge density. Size measurements typically are performed by dynamic light scattering. Here we focus on the measurement of liposome charge by the technique known as particle electrophoresis or microelectrophoresis. As will be shown, the electrophoretic charge measurement is dependent on the liposome size and surface structure.

Particle electrophoresis differs from gel electrophoresis in that the particles of interest are suspended in free liquid, which in our case is water or aqueous electrolyte. There is no external resistance to the movement of the particles except that due to the electrolyte itself. The force driving the motion of a particle is QE, where Q is the net electrical charge of the particle, and E is an externally applied electric field. The electrical force accelerates the particle. However, because of several retarding forces including friction, the particle rapidly attains a terminal velocity known as its electrophoretic velocity. The magnitude of this velocity is, of course, related to the driving force QE. Because E is known from the experimentally applied voltage, measurement of the velocity yields information about Q.[1] This technique is a powerful method for determining the net electrical charge of liposomes.

Later we outline the procedure for calculating the charge from the observed velocity. However, a qualitative result, requiring no calculations, is first worth noting. The *sign* of the net particle charge is given immediately by the direction in which the particles move. If they move toward the positive electrode, then they are net negatively charged, and vice versa. For many applications, this information is sufficient. When several different kinds of particles of the same size and surface structure are present in the same electrolyte, the faster ones are the more highly charged. Stationary particles have zero net charge.

---

[1] For details, see R. J. Hunter, "Zeta Potential in Colloid Science: Principles and Applications." Academic Press, London, 1981.

It is important to note that the "particle" in this discussion is a hydrodynamic particle, which is not necessarily the same as the physical particle. An electrophoretically driven particle, even if perfectly smooth, cannot shear the first layer (or layers) of water at its surface. This phenomenon is known as the no-slip boundary condition. Therefore some water is dragged along with the particle. The particle-associated water extends out to a so-called shear surface, where shearing of the water phase begins. The hydrodynamic particle includes everything encompassed within the bounds of the shear surface. With regard to charge, the hydrodynamic liposome includes the intrinsic lipid headgroup charge; the charge of any covalently attached, adsorbed, or trapped moieties; and electrolyte screening charge located both within the liposome and extending out as far as the shear surface. Typical shear surface distances for common liposomes have been estimated to be $\sim$2 Å in 100 mM NaCl, 4 Å in 10 mM NaCl, and 10 Å in 1 mM NaCl,[2] although for liposomes with "fuzzy" surfaces the effective shear surface distances can be much larger.[3] A considerable amount of screening charge can be contained in the region between the physical liposome surface and the shear surface, especially for highly charged liposomes at high ionic strength, and especially if multivalent counterions are present. Because the experimenter usually is interested in charge associated with the physical liposome, the screening charge located between the physical surface and shear surface must be determined and subtracted from the total electrophoretically measured charge. This computation can be accomplished readily by use of theoretical formulas if the distance of the shear surface from the physical surface of the liposome is known or assumed.[1,3,4] A lack of precise knowledge about the shear-surface location is perhaps the main difficulty in the quantitative interpretation of electrophoretic measurements.

Equipment: Rank Brothers Apparatus

The experimentally determined quantity is the electrophoretic velocity. The simplest and most straightforward device for this measurement, in our experience, is the Rank Brothers Mark II particle microelectrophoresis apparatus (Rank Brothers, Bottisham, UK). A description of the operation

[2] R. V. McDaniel, A. McLaughlin, A. P. Winiski, M. Eisenberg, and S. McLaughlin, *Biochemistry* **23**, 4618 (1984).
[3] J. A. Cohen and V. A. Khorosheva, *Colloids Surf. A Physicochem. Eng. Aspects* **195**, 113 (2001).
[4] J. A. Cohen, B. Gabriel, J. Teissié, and M. Winterhalter, *in* "Planar Lipid Bilayers (BLMs) and Their Applications" (H. T. Tien and A. Ottova, eds.), p. 847. Elsevier Science, Amsterdam, 2003.

of this apparatus demonstrates the essential features of electrophoretic measurements in general. These principles are also relevant to the more modern instruments.

Cell and Electrodes

In the Rank Brothers apparatus, the liposome suspension is enclosed in a thin-walled glass cell of either cylindrical or flat geometry. The flat cell, shown in Fig. 1, is by far the simpler to use. The electrophoretic chamber consists of two vertically oriented, hydrodynamically smooth, parallel silica faces 40 mm long × 2.5 mm high, spaced 0.5 mm apart. The gap between the faces is closed at the top and bottom, forming a long, tall, narrow channel of rectangular cross-section, shown in Fig. 2. Variously sized cells are available. There is a significant advantage to a cell whose major hydrodynamic surfaces are oriented vertically, rather than horizontally. With vertical walls, sedimenting particles do not settle on a hydrodynamically important surface, thus do not perturb electroosmotic flow in the cell upon which accurate measurements depend. On both open ends of the rectangular channel the Rank Brothers cell widens to form arms that bend upward and terminate in tapered ground-glass sockets. The channel volume is 50 μl. The volume of the entire cell is about 10 ml.

The cell may have either two or four electrodes. The two-electrode cell is more convenient for semiquantitative measurements, whereas the four-electrode cell (see Fig. 1) is recommended for highest accuracy. The inner "voltage" electrodes are two platinum wires that protrude into the cell

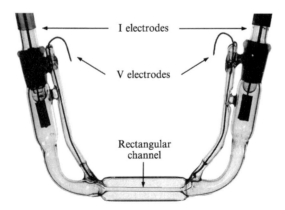

I electrodes

V electrodes

Rectangular
channel

Fig. 1. Rank Brothers four-probe flat cell. The electrophoresis channel is 40 mm long × 2.5 mm high × 0.5 mm deep. The channel volume is 50 μl and the entire cell volume is ~10 ml.

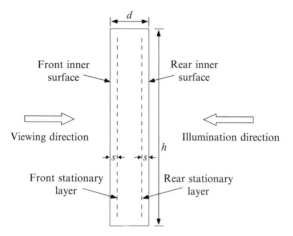

Fɪɢ. 2. Transverse cross-section of the cell channel, drawn to scale. For electrophoretic measurements, the microscope focal plane is positioned at either the front or rear stationary layer. When the electric field is applied, the liposomes move either into or out of the page in this representation. For the cell described in text, $h = 2.5$ mm, $d = 500\,\mu$m, $s = 88.6\,\mu$m.

through glass seals near the open ends of the rectangular channel, and then pass upward through glass supports and emerge at the top of the cell, where they are connected to a high-impedance voltmeter (6512 digital electrometer; Keithley, Cleveland, OH). The outer "current" electrodes are made of platinum foil bent into cylinders and soldered to platinum wires embedded in removable glass holders. The holders terminate in tapered ground-glass plugs that fit into the ground-glass sockets of the cell. The wires protruding from the tops of the glass holders are connected to the Rank Brothers machine, which controls and monitors the cell current and also monitors the voltage across the outer electrodes.

Before placing the electrode holders in the cell sockets, the cell is filled carefully with the liposome-containing solution, avoiding the introduction of bubbles. The cell is filled to the top so that some solution overflows when the holders are inserted, ensuring that no air space remains in the closed cell. The holders are inserted smoothly to avoid formation of bubbles and are seated firmly with a slight twist, using two hands to avoid undue torque on the cell. Placement of the second holder may dislodge the first holder, so both must be rechecked for tightness. To avoid contamination of the liposome-containing solution, no sealing grease is used on the ground-glass joints. When the holders are seated properly, the cell is sealed airtight and the current electrodes are positioned in the arms of the cell (see Fig. 1).

The power supply can be operated in either a constant-voltage or constant-current mode. For the four-probe cell, the constant-current source is preferred. Current is delivered via the two large outer electrodes. Electrolyte resistance in the channel creates an *IR* voltage drop across the cell, which establishes the electric field ($E = V/\ell$, where V is the voltage measured between the two inner electrodes and ℓ is the effective distance between these electrodes; see Calibrations later). The utility of this arrangement is that, because of the high impedance of the electrometer voltmeter, only a minuscule amount of current is drawn through the voltage electrodes, producing a negligible *IR* voltage drop across the electrode–electrolyte interfaces. Thus, the monitored voltage accurately measures the voltage drop across the channel, hence gives the true voltage gradient (i.e., electric field) "seen" by liposomes in the cell. The voltage difference between the outer current electrodes is somewhat variable and is typically 10–20% higher than that across the inner voltage electrodes, because it includes *IR* voltage drops across the electrode–electrolyte interfaces, which are not sensed by the liposomes. During a run, electrode polarization or electrolytic bubble formation may increase the interfacial resistance of the current electrodes. The constant-current source counteracts this effect by boosting the applied voltage to maintain a constant current. Thus the voltage read by the inner electrodes remains nearly constant. In any case, we record the voltage across the inner electrodes for each velocity reading, which gives an accurate measure of the field sensed by the liposomes during the time of the measurement. Current is monitored with a digital ammeter.

Optics

The cell is illuminated from behind by a focused quartz-iodide lamp providing dark-field illumination. Liposomes in the cell, if large enough, appear as bright spots visible through the front cell wall via a horizontal-viewing microscope. The microscope is mounted on a spring-loaded horizontal stage that can be moved toward or away from the cell by use of a precision mechanical micrometer. The microscope has a sharp focal plane that can be positioned back and forth inside the cell by turning the micrometer knob. The cell itself is mounted on a vertical stage, and another precision micrometer can move it up and down. The mechanical micrometers are of very high quality (L. S. Starrett, Skipton, UK). Positioning is reproducible with 1-μm precision.

The microscope eyepiece contains a custom-designed grid of 0.15-mm squares, accentuated at every tenth line, with bold horizontal and vertical midlines (Graticules, Tonbridge, UK). The eyepiece graticule is used to measure distances inside the electrophoresis cell and is calibrated in two ways. First, a glass micrometer standard having accurate 10-μm divisions (model S22; Graticules) is mounted vertically in the water-filled bath approximately where the electrophoresis cell will be placed. The microscope is focused on the micrometer scale, and the eyepiece grid is compared with this scale, yielding the lateral distance in the bath corresponding to one square of the eyepiece grid. A second method relies on the accuracy of the mechanical micrometers. The electrophoresis cell is filled with water and mounted in the bath, then the microscope is focused on a spot or scratch on the front inner surface of the cell channel. (Later this procedure is repeated with the rear inner surface and the results are averaged.) The vertical micrometer is used to move the cell vertically, sweeping the spot from top to bottom of the eyepiece grid. The actual distance traveled is read on the mechanical micrometer scale, yielding the distance traveled per square of the eyepiece grid. We find that the two methods give comparable results, which also confirms the accuracy of the mechanical micrometers. In our case, each eyepiece grid square corresponds to a distance of about 8 μm inside the electrophoresis channel. Both methods give actual transverse distances, unperturbed by the indices of refraction of the water and glass. [When in air looking through a planar air–water interface at an object in the water, the apparent transverse dimensions of the object are not altered, but the apparent distance of the object from the interface is smaller than its actual distance by a factor $n_{air}/n_{water} \approx 0.75$. These statements are true when the angles of incidence and refraction at the interface are small (paraxial approximation, i.e., angles $\lesssim 10°$).[5]]

Temperature Control

Electrophoretic measurements conventionally are performed at 25°. The electrophoresis cell is mounted in a metal holder that clamps to the wall of a Plexiglas tank, which serves as a thermal water bath. The bath temperature is controlled to $\pm 0.02°$ by a submerged heater driven by a proportional temperature controller (YSI 72; Yellow Springs Instruments, Yellow Springs, OH), which in turn is driven by a shielded thermistor (YSI 406) positioned in the bath near the cell. The bath water is circulated via an external pump through closed-circuit Tygon tubing passing through a crushed-ice reservoir. The proportional controller holds the temperature extremely well at its nominal set point; however, we find that the set point is not accurate. A precision glass–mercury thermometer, marked in 0.1° divisions between 20 and 30° (model 21256; Brooklyn Thermometer,

[5] F. W. Sears and M. W. Zemansky, "University Physics," 3rd Ed., p. 864. Addison-Wesley, Reading, MA, 1964.

Farmingdale, NY) is used to monitor the bath temperature. The controller set point is then adjusted to yield a bath temperature of $25.00 \pm 0.05°$ according to the thermometer. Because electrophoretic velocities are temperature dependent, it is important that this parameter be well controlled.

It should be appreciated that the steady-state temperature of the electrolyte in the cell is not necessarily the same as the temperature of the bath, especially if high currents are applied, which can cause ohmic heating. Electrolyte heating may also induce convection in the cell, which can perturb the stationary layers (see later). A good measure of the heating of an electrolyte is its electrical conductivity, which is sensitive to temperature and is monitored easily via the cell conductance $= I/V$. (Conductance $=$ conductivity $\times A/\ell$, where A is the cross-sectional area of the channel and ℓ is the effective distance between the voltage electrodes.) Electrolyte conductivity rises typically about 2% per degree Celsius. Thus if the conductance is monitored to three significant figures, a rise on the order of $0.1°$ can be detected. If such changes are observed, the applied current must be reduced. To minimize heating, electrophoretic readings are taken as quickly as is practical, the current is turned off between readings, and at least several minutes are allowed between consecutive runs. We find that maintaining the applied power ($P = IV$) below 20 mW (e.g., 1 mA at 20 V) is generally sufficient to prevent detectable heating of the electrolyte. It is also prudent to turn off the quartz-iodide lamp between runs to minimize cell heating by infrared absorption. The lamp life is prolonged by turning the lamp on and off gradually with a rheostat.

Calibrations

Stationary Layers

Because the glass walls of the cell channel are intrinsically charged (due to dissociated silicate groups), the applied electric field induces electroosmotic flow in the channel. The reason is that a diffuse screening layer of cations forms in the electrolyte near each negatively charged wall. These ions are driven by the applied field, collide with water molecules, and induce bulk water flow along the walls. Because the cell is closed, water moving in one direction along the walls must return in the opposite direction via the center of the cell. The velocity profile through the cell is parabolic, with water flowing in a positive direction adjacent to the vertical walls, and in a negative direction in the vertical plane midway between the walls. ("Positive" here is defined as the direction in which positive ions move, i.e., toward the negative electrode.) There are two vertical planes in the channel, known as the stationary layers, where the water velocity changes from positive to negative, thus is zero. In these two planes there is no electroosmotic flow, and the bulk water is stationary. Electrophoretic measurements must be made at these stationary layers, because any particle velocity there is driven by QE alone, unaffected by electroosmotic convection.

Interestingly, the contribution of electroosmotic convection to the observed velocity of a particle cannot be averaged out by reversing the applied electric field. Reversal of the field reverses the directions of both the electroosmotic flow and the velocity of the particle; hence any electroosmotic contribution to the particle velocity remains unchanged. The only ways to overcome electroosmotic convective effects are (1) to measure at the stationary layers, (2) to chemically neutralize the glass surfaces by treatments such as silanization, or (3) to use an Uzgiris-type cell, in which flat electrodes are positioned in the middle of the cell far from the walls.

We find method (1) to be relatively straightforward. It requires that the stationary layer locations be determined accurately, which may be done both experimentally and theoretically.

Experimental Determination. Electrophoretic measurements are performed with liposomes made of fresh egg phosphatidylcholine (PC) (Avanti Polar Lipids, Alabaster, AL) in 10 mM NaCl electrolyte. PC is zwitterionic near pH 7 and thus has no net charge, so the electrophoretic velocity of PC liposomes is zero. Therefore any observed liposome velocity is due to convective flow. The velocity profile of the water in the channel is plotted using the horizontal stage micrometer to step the microscope focal plane through the channel in increments equal to 10% of the channel thickness. In this procedure, all distances are apparent distances. Measurements are made at a height midway between the top and bottom of the channel. The first reading is taken as close to the inner surface of the front wall as possible, and the last at the inner surface of the rear wall. Liposomes near the walls that appear stationary are undoubtedly stuck to the walls and should be ignored. (The no-slip regions at the wall surfaces are too narrow to be observed.) Eleven readings are taken, with one at the cell center. A parabolic fit of velocity versus distance should be symmetric about the channel center. If it is not, the cell requires refilling to eliminate bubbles or liposomes on the walls and/or needs cleaning (see later). The distances from the inner surfaces where the velocity is zero are determined from the parabolic fit. These distances are the apparent locations of the stationary layers, s_{app}. The values should agree with those calculated from Eq. (1) on p. 156.

Theoretical Determination. The Komagata formula is used.[6,7] Let s be the distance of the front stationary layer from the front inner surface, which equals the distance of the rear stationary layer from the rear inner surface; d is the depth or thickness of the cell, that is, the distance between the inner surfaces of the front and rear walls; and h is the height of the cell, that is, the distance between the top and bottom inner cell surfaces (see Fig. 2). The Komagata formula for the flat cell is

$$\frac{s_{app}}{d_{app}} = \frac{s}{d} = \frac{1}{2} - \left(\frac{1}{12} + \frac{32}{\pi^5}\frac{d}{h}\right)^{1/2} = 0.5 - \left(0.0833 + 0.1046\frac{d}{h}\right)^{1/2} \quad (1)$$

where s, d, and h are actual distances, and s_{app} and d_{app} are apparent distances when the cell is filled with water or electrolyte. The relations between the apparent and actual distances are $s_{app} = s/n_w$ and $d_{app} = d/n_w$, where n_w is the index of refraction of water, which is 1.3325 at 25°. The distances d and h can be measured accurately by use of the microscope and mechanical micrometers. With air in the cell, the microscope is first focused on scratches or spots on the front inner surface approximately midway between the top and bottom of the channel. The horizontal micrometer knob is then turned steadily (in one direction only, to avoid backlash) until scratches or spots on the rear inner surface come into focus. The micrometer scale gives the distance d traveled, accurate to $\pm 1\ \mu m$. We do such measurements three times and take the average.

The interior height of the cell is measured by the vertical micrometer. The focal plane of the microscope is first set at the cell midplane with the horizontal micrometer. The cell is then moved vertically with the vertical micrometer until the bottom inner surface of the cell coincides with the horizontal midline ("equator") of the eyepiece grid. The vertical micrometer knob is then turned steadily in one direction until the top inner surface lies on the grid equator. The micrometer distance traveled $= h$.

Once d/h is known, the ratio s/d may be calculated from Eq. (1). It ranges from 0.2113 when $d \ll h$ (cell thickness much smaller than its height, so that the cell approximates two infinite parallel planes) to 0.0665 when $d = h$ (cell has a square cross-section). When d is comparable to h, s/d depends on the vertical position in the cell. The Komagata formula gives s/d at a height midway between the cell top and bottom, which is where the measurements should be made. In Fig. 2 where $d/h = 0.2$, Eq. (1) gives $s/d = 0.1772$.

[6] S. Komagata, *Res. Electrotech. Lab. (Tokyo)* **348,** 1 (1933).
[7] H. B. Bull, "An Introduction to Physical Biochemistry," 2nd Ed., p. 332. F. A. Davis, Philadelphia, 1971.

Because experiments are of course performed with water in the cell, s_{app} is needed. It is found via $s_{app} = s/n_w$, or equivalently by $s_{app} = (s/d)d_{app}$, where (s/d) is obtained from the Komagata formula, and d_{app} is found by measuring the cell thickness with electrolyte in the cell. Measurement of d both in the absence and presence of electrolyte in fact gives the index of refraction of the electrolyte. This is more accurate than assuming n_w, especially for concentrated electrolytes where the index of refraction may be larger than that of water, although for most experiments the difference is negligible.

Before using a new cell, it is recommended that s_{app} be determined experimentally and shown to yield the same result as the Komagata formula. If the two do not agree, then the cell is not reliable. Once agreement is demonstrated, the experimental procedure need not be repeated, and the formula may be used routinely to find s_{app}.

The parameter s_{app} is used for measurements as follows. First, the precise locations of the front and rear inner surfaces of the channel must be found. The cell position changes significantly (much more than $\pm 1\,\mu m$) whenever the cell is removed from the apparatus for cleaning or reloading. It is useful to have drawn a "map" of scratches and spots on the front and rear inner surfaces of the channel as they appear through the microscope when the cell height is adjusted such that the eyepiece graticule equator lies midway between the top and bottom of the channel. We use graph paper, with each square corresponding to a square on the eyepiece grid. In this manner the inner surfaces are easily recognized, and the proper vertical position of the cell can be established quickly and accurately. Starting with the focal plane inside the front glass wall of the channel, the horizontal micrometer position is advanced and its position noted when the front surface landmarks come into focus. The micrometer is then advanced (by turning clockwise) a further distance $= s_{app}$, which puts the microscope focal plane at the front stationary layer. Any liposomes now in focus are located within $\sim 1\,\mu m$ of this stationary layer and can be used for electrophoretic measurements. The rear inner cell surface is located similarly, and the micrometer is backed off (by turning counterclockwise) a distance $= s_{app}$, which places the microscope focal plane at the rear stationary layer. [When backing off the micrometer (turning counterclockwise), it should be turned well past the end point and then advanced (turning clockwise) up to the end point. Thus, the end point is always approached by turning the micrometer in the same direction, which eliminates errors due to micrometer backlash.] Electrophoretic measurements are made at both stationary layers and compared (see later).

Cell Constant

The cell constant is the effective distance ℓ between the inner voltage electrodes to be used in calculating the electric field strength from the measured voltage difference between these electrodes ($E = V/\ell$). If the electrodes were large plates separated by a uniform conducting channel, the effective distance would be the actual distance between the plates. The parameter ℓ is calculated from the formula $G = K(A/\ell)$, where G is the measured cell conductance when the cell is filled with a conductivity standard (an electrolyte of known conductivity K), and A is the cross-sectional area of the channel ($A = hd$), where d is the actual (not the apparent) thickness of the channel. Thus, $\ell = A(K/G)$.

For the conductance measurements, the cell must be cleaned (see below) then filled with a standard electrolyte, avoiding the presence of bubbles and dilution by residual water. The cell is rinsed twice with the electrolyte to be used in the measurement, and then the electrolyte is allowed to sit in the cell overnight. We use a KCl conductivity standard close to the ionic strength of interest, that is, 0.1000 or 0.0100 M (Ricca Chemical, Arlington, TX; Orion Research, Beverly, MA). The current electrodes are inserted (to approximate the cell geometry of an electrophoretic measurement), the cell is placed in the water bath, and the temperature is allowed to equilibrate for 1 h to $25.00 \pm 0.02°$. Conductance is now measured across the inner voltage electrodes with an RC-20 low-frequency conductance bridge (Altex-Beckman, Cedar Grove, NJ) with fresh batteries. When performing this measurement, all electrical leads to the cell except those to the bridge must be disconnected. Readings are repeated until the conductance stabilizes, indicating that the temperature in the cell has reached a steady state. The cell is then emptied and reloaded, and the measurement is repeated three times, as it is always possible that bubbles might be trapped in the cell, yielding artificially low conductances. Repeated measurements should be reproducible to several percent. A good test of the reliability of the method and the standards is to take the ratio of the measured conductances with two separate conductivity standards, for example, 0.1 M KCl and 0.01 M KCl. Because the cell geometry is constant, this ratio should agree with the theoretical conductivity ratio. At $25°$ the theoretical conductivity of 0.1 M KCl is 12,896 μS/cm, and that of 0.01 M KCl is 1412.7 μS/cm; thus the ratio is 9.1286. (S is "siemens" $\equiv$ 1/ohm, formerly called "mho.")

Once G has been measured, ℓ is calculated from G, K, and A according to the previously-described formula. The values obtained for ℓ should be independent of the electrolyte (i.e., the ratio K/G depends only on the cell geometry). In our case, ℓ is about 4 cm, which is close to the physical distance between the voltage electrodes at the two ends of the electrophoretic channel.

Electrode Polarity

The current applied to the current electrodes is controlled by an on/off and reversing switch that allows the current to be applied alternately in each direction. It is necessary to establish which electrophoretic direction is positive and which is negative as viewed through the microscope eyepiece. This may not be obvious, because the optics can invert the visual field. One switch position is labeled "normal" and the other "reverse." The current electrode holders are also labeled; for example, one is colored red and one black. A control experiment is performed with liposomes of known charge polarity, for example, phosphatidylserine⁻. The electrodes then are placed so that, with the switch in the normal position, the red electrode is positive, the apparent movement of negatively charged particles at the stationary layer is to the left, and that of positive particles is to the right. Thereafter, the electrodes are always placed in the same positions, so that the sign of the particle charge is clearly evident.

Methods

Solutions

Water is pretreated, then purified, deionized to resistivity > 18 MΩ-cm, and postfiltered through 0.2-μm filters (E-pure; Barnstead, Dubuque, IA). NaCl solutions of various ionic strengths are diluted from a stock solution of 100 mM NaCl, 10 mM 3-(N-morpholino) propanesulfonic acid (MOPS) buffer, and 1 mM ethylenediaminetetraacetic acid (EDTA), pH 7.2–7.6. Acidic buffers are preferable for use with negatively charged liposomes to minimize electrostatic interaction of the buffer with the liposome surface charge. For example, basic Tris buffers have been shown to bind to negatively charged phosphatidylserine liposomes and to reduce their net surface charge density.[8] The function of EDTA is to chelate contaminant multivalent cations in the electrolyte, which can bind strongly to liposomes and alter their net charge. For studies with added multivalent cations in which EDTA is omitted, controls are first performed without the added multivalents, comparing the liposome mobilities in the presence and absence of

---

[8] M. Eisenberg, T. Gresalfi, T. Riccio, and S. McLaughlin, *Biochemistry* **18,** 5213 (1979).

EDTA. This procedure checks whether the EDTA-free solution is free of contaminating multivalent cations.

The pH of stock solutions is adjusted with $1.000 \pm 0.001 \, N$ NaOH made with CO_2-free water and traceable to NIST (National Institute of Standards and Technology) standards (Ricca Chemical). To remove multivalent cation impurities, the NaOH is filtered twice through 25-mm Bio-Rex sample preparation disks containing Chelex ion-exchange membranes (Bio-Rad Laboratories, Hercules, CA). A syringe is Luer-locked to the preparation disk, and the NaOH is pushed slowly through the disk under pressure. Contact between the plunger tip and the strong NaOH is minimized by keeping an air space between them. It should be noted that the filters are unidirectional and must be prerinsed.

In titrating the stock solutions, it is necessary to control and record the total amount of NaOH added, because free Na^+ affects the ionic strength of the final electrolyte. It is best to avoid overshooting the desired pH, then back-titrating with HCl, because all added ions increase the ionic strength. To make the EDTA solution, 50 mmol of 99% pure disodium EDTA dihydrate (SigmaUltra grade; Sigma, St. Louis, MO) is dissolved in about 400 ml of deionized water and titrated with 1-ml increments of the filtered $1 \, M$ NaOH while stirring and monitoring pH. We use a solid-state pH electrode (IQ Scientific Instruments, San Diego, CA). The amount of NaOH added is typically 46.7 mmol. Making up the final volume with deionized water yields 500 ml of $0.1 \, M$ $Na_{2.93}$ EDTA at pH 7.2. The solution is then filtered through 0.2-μm Nuclepore filters (Nuclepore Filtration Products, Pleasanton, CA) and stored in a refrigerator.

For the NaCl–MOPS–EDTA stock solution, we start with 50 mmol of ACS-certified NaCl crystals, 50 mmol of 99.5% pure MOPS free acid (SigmaUltra grade), plus 10 ml of the previously-described $0.1 \, M$ EDTA stock solution in about 850 ml of deionized water. Filtered $1 \, M$ NaOH is added in 1-ml increments while stirring and monitoring the pH. Although the pK_a of MOPS at 25° is 7.2, the solution is titrated up to pH 7.6, because electrophoretic measurements tend to reduce the pH. Typically 8 mmol of NaOH is required. Making up the final volume with deionized water yields 1 L of nominal 100 mM NaCl, 10 mM MOPS, 1 mM EDTA, pH 7.6. Counting the 10.93 mmol of Na^+ added with the disodium EDTA and NaOH, total $[Na^+]$ is 111 mM. Accounting for the $EDTA^{3-}$ and $EDTA^{2-}$ moieties yields an ionic strength of 114 mM.

The stock solution is diluted to the ionic strength desired for the experiment. For electrophoretic determination of liposome charge density, low ionic strengths are advantageous. High ionic strength causes significant counterion screening of the liposome charge behind the shear surface (i.e., within the hydrodynamic particle), and thus reduces the signal. For this reason electrophoretic measurements of weakly charged liposomes usually are not conducted at physiologic ionic strengths ($\sim$100–150 mM). On the other hand, extremely low electrolyte concentrations may not provide sufficient buffering capacity to control the pH. A reasonable compromise is 10 mM electrolyte with 1 mM pH buffer.

Whenever the stock is diluted, its pH is rechecked. If [MOPS] is diluted below 1 mM, it may not provide sufficient buffering capacity during an experiment. In this case it is essential to check the pH of the liposome-containing solution both before and after a measurement. It should be appreciated that the electrolyte pH "seen" by protonatable groups on the liposome surface is the interfacial pH, called pH$_0$, which differs from the bulk pH if the surface is charged. Protons are electrostatically attracted to, or repelled from, the surface depending on whether the surface is negatively or positively charged, respectively. From the Boltzmann relation,

$$[H^+]_0 = [H^+]\exp\left(\frac{-e\psi_0}{kT}\right) \quad \text{or} \quad pH_0 = pH + \frac{e\psi_0}{2.303kT} \qquad (2a)$$

At 25°,

$$pH_0 = pH + \frac{\psi_0}{59.16} \qquad (2b)$$

where "pH" means the bulk pH as measured with a pH electrode, and the electrostatic surface potential ψ_0 is in units of millivolts. Thus if $\psi_0 \simeq -59$ mV, the electrolyte pH adjacent to the liposome surface is 1 pH unit below that in the bulk. This effect cannot be overcome by the bulk-phase pH buffer. Determination of ψ_0 by electrophoretic measurements is described later.

If bubbles are found to form in the electrophoresis cell, it is useful to degas the deionized water before making up the solutions. This is done by pumping on the water briefly in a vacuum desiccator using a liquid N$_2$ cold trap. Because pumping cools the water, the water should first be heated slightly, ensuring that it is degassed at 25°. The degassed water is stored in an airtight container.

Liposome Preparation

Phospholipids are purchased in chloroform solution from Avanti Polar Lipids. The lipid–chloroform solutions are removed from their sealed ampules and stored in 5-ml conical Teflon–PFA vials with threaded Teflon caps (model 024; Savillex, Minnetonka, MN). Solutions received in glass bottles with Teflon caps may be stored in the bottles after opening, but a brief purge with high-purity N_2 or argon gas is recommended to retard oxidation of unsaturated lipids. A problem with Teflon bottle caps is that they may gradually loosen due to the high chloroform vapor pressure. Hard plastic caps with Teflon liners may be used, but the liners must be held in place by a press-fit, not by adhesives. Any additional sealing of vessels is done with Teflon tape, not Parafilm, which chloroform dissolves. The lipid–chloroform solutions are stored at $-20°$. Before a bottle or vial is opened, it should be warmed to room temperature to prevent condensation from the air. Solutions may be cloudy or gelled when cold, but should be completely clear and homogeneous before dispensing.

To make liposomes composed of several phospholipid species, it is first necessary to know the lipid concentration of the lipid–chloroform solution of each component. Lipid–chloroform solutions obtained from Avanti may vary by $\sim$5–10% from their nominal values (usually they are more concentrated than nominal). Also, the lipid concentrations in the Savillex vials increase because of chloroform evaporation whenever the vials are opened. This problem is minimized by keeping the vials and microdispensers chilled, opening and closing the vials quickly, doing transfers in a cold room, and using moderately concentrated lipid–chloroform solutions to reduce the need for multiple pipetting. We use a lipid–chloroform concentration of $\sim$70 mg/ml for the major component of a lipid mixture (in our case phosphatidylcholine) and $\sim$20 mg/ml for the minor component. The lipid concentration may be altered for convenience. It can be decreased by adding high-purity chloroform with a glass-bore microdispenser (see below), or it can be increased by carefully blowing off chloroform from the solution with high-purity dry N_2 or argon gas in a fume hood.

Chloroform is HPLC grade and must be handled with care. It is kept in a fume hood either in glass or in Nalgene Teflon (FEP) flexible wash bottles with Tefzel (ETFE) tubing and screw closures (Nalge Nunc International, Rochester, NY). It is important not to store chloroform in, or expose it to, any plastic that it attacks, which includes virtually all plastics except Teflon. Breathing chloroform vapor and exposure to skin should be avoided.

The procedure for making mixed component liposomes is as follows.

1. The weight-to-volume (mg/ml) concentration of each component lipid in its chloroform solution is measured gravimetrically. By use of an adjustable digital microdispenser with glass bore and positive-displacement Teflon-tipped plunger (VWR Scientific, San Francisco, CA), $20.0 \,\mu$l of lipid–chloroform solution is dispensed into a small aluminum pan on a hot plate. (Conventional pipettors with polypropylene tips cannot be used with chloroform.) After heating for 2 min to evaporate the chloroform, the

pan plus dried lipid is weighed on a Cahn C-31 digital microbalance (Thermo Orion, Beverly, MA) to three significant figures (1-μg precision). The weight is recorded 30 s after placing the pan on the balance, as measured by a digital timer. The pan is then cleaned with chloroform, first by using a squeeze bottle (see above) and then by submerging it in chloroform in a small beaker. It is then dried on the heater, weighed, tared, and the procedure repeated. The average and standard error of the mean (SEM) for three measurements are calculated. If necessary, measurements are repeated until the SEM is less than 2% of the mean value. We find that the measured weights of phosphatidylcholine (PC) have significant scatter, whereas the weights of phosphatidylserine (PS) and phosphatidylglycerol (PG) are quite reproducible. The averaged lipid weight divided by 20.0 μl gives the weight-to-volume concentration of the lipid–chloroform solution.

For quality control the lipid concentration of each lipid–chloroform solution is remeasured after every three film preparations. Typically the lipid concentrations increase by $\sim$2% from one concentration measurement to the next. By frequent corrections for the changing lipid concentrations, the compositions of the lipid mixtures can be controlled accurately. We estimate the molar ratios of our lipid mixtures to be accurate to $\pm$2%. Ben-Tal et al.[9] report that gravimetric determinations of lipid concentrations give the same results as phosphate analysis.

2. To make lipid mixtures at the desired molar ratios, the correct volume of each lipid–chloroform solution is calculated from its weight-to-volume concentration and the molecular weight of the lipid, such that total lipids add up to 20 μmol. The proper aliquots are then dispensed into a 25-ml round-bottom glass flask, using the VWR positive-displacement adjustable digital microdispensers.

3. The flask is shaken briefly to mix the chloroform solution thoroughly, then is subjected to rotary evaporation, using a mechanical vacuum pump with a liquid N_2 cold trap (model 9466R75; Thomas Scientific, Swedesboro, NJ). The flask is kept under vacuum for at least 1 h to dry the lipid film. If the film is not used right away, the flask may be purged with high-purity N_2 or argon gas, sealed, and kept in a refrigerator for up to several days without noticeable degradation of the film.

4. The aqueous buffer to be used in the experiment is added to the flask to give 2 ml of 10 mM lipid. The mixture is gently swirled around until the dried lipid film dissolves off the glass walls, producing a turbid suspension of polydisperse multilamellar vesicles. The suspension can be vortexed briefly, although in some cases this makes the liposomes too small to be

[9] N. Ben-Tal, B. Honig, R. M. Peitzsch, G. Denisov, and S. McLaughlin, Biophys. J. **71**, 561 (1996).

seen in the Rank Brothers machine ($\lesssim 1 \mu$m diameter). Hydrating should be done at a temperature above the gel–liquid crystalline phase transition temperatures of all the component lipids. Unless used right away, the flask is purged with N_2 or argon and sealed.

5. If a monodisperse suspension of unilamellar liposomes is desired, the suspension is cycled through five freeze–thaw cycles to disrupt and hydrate the lamellae. Bio-freeze vials (5 ml; Costar, Cambridge, MA) containing the liposome suspension are lowered into liquid N_2 and thawed in warm water. The suspension is then loaded into a 1-ml glass gas-tight syringe and passed 11 times through a polycarbonate membrane having the desired pore size, typically 100 or 200 nm, using an extrusion device (Avanti Polar Lipids; or Avestin, Ottawa, ON, Canada). The extrusion must be done at a temperature above the phase-transition temperatures of all the component lipids. Extruded liposomes are too small to be seen with the Rank Brothers machine.

6. Finally, ~ 0.4 ml of liposome suspension is diluted in 10 ml of electrolyte buffer and added to the electrophoresis cell for electrophoretic measurements. The solutions used to hydrate the lipids and to dilute the liposomes should be the same.

Blackening Electrodes

The current and voltage electrodes can be made more stable, with decreased interfacial resistance and less tendency to polarize, by "blackening" them. Blackening is an electrolytic process that deposits a porous platinum coat on the electrode surface, which greatly increases its surface area.

The electrodes are first cleaned by soaking for 5 min in aqua regia (82% HCl, 18% HNO_3, v/v) and then for 5 min in hot detergent at 50°. The detergent is RBS 35 (Pierce Chemical, Rockford, IL), diluted to 20 ml/L in deionized water. The electrodes are then rinsed 10 times in deionized water. It is important to remove all detergent to prevent its later introduction into the electrophoresis cell.

Fifty milliliters of electroplating solution is made and stored for repeated use. It consists of 2% (1 g/50 ml) chloroplatinic acid ($H_2PtCl_6 \cdot 6H_2O$), also called "platinic chloride" (P-5775; Sigma) and 0.02% (10 mg/50 ml) lead acetate ($PbAc_2 \cdot 3H_2O$) [L-3396 (Sigma) or I-2771 (J. T. Baker, Phillipsburg, NJ)]. When new, the solution is opaque orange, but after many uses it gradually becomes a clear pale yellow, at which point it should be safely discarded.

Electroplating is done with a 6-V motorcycle battery trickle charger (Schumacher Electric, Mount Prospect, IL). A platinum wire is used as the anode (positive terminal), and the electrode to be blackened is the cathode. A 10-kΩ variable resistor and DC ammeter are used to limit and monitor the current, respectively. The large current electrodes are plated, one at a time, in a small beaker with 50 ml of plating solution. The applied current is set initially at 15–20 mA. Bubbles may appear on the platinum wire anode, but no bubbles should be allowed to form on the electrode being blackened. The electroplating solution should not be stirred. The electrode becomes gray, then a velvety black, within 5 min. If it does not, the current is increased gradually to 30 mA. The electrode is then rinsed thoroughly in deionized water and dried in air. For the small voltage electrodes in the electrophoresis cell, the same procedure is repeated, except that the plating solution is poured into the cell and the platinum-wire anode is inserted into one of the arms of the cell. The electrodes are plated one at a time with a plating current of 1 mA. The plating is completed in ~10 s. The plating solution is then recovered and the cell is rinsed thoroughly.

Replating is necessary when the current electrodes become scratched and gray or shiny, which is about every few months with constant use. In our experience, the inner voltage electrodes need to be blackened only once.

Cleaning Cell, Glassware, and Electrodes

Careful cleaning of the cell, electrodes, and glassware is important for achieving reproducible results. After use, the cell is first rinsed with deionized water, acetone (to remove the water), 2:1 (v/v) chloroform–methanol (which is miscible with acetone and dissolves lipids), acetone, and then water. (Acetone is used as a solvent "bridge" between water and chloroform–methanol, which are immiscible.) All organic solvents are high-purity grade. The cleaning solution is made from Chromerge (Fisher Scientific, Pittsburgh, PA). Chromerge is added to concentrated sulfuric acid at a 1:100 (v/v) ratio, producing a saturated chromic/sulfuric acid solution, which is then diminished to half strength by 1:1 (v/v) dilution in water. (For safety, the acid is added slowly to the water, not vice-versa.) The cell is filled with the solution, allowed to sit for 5 min, then emptied and rinsed with water. The orange-brown cleaning solution can be saved and reused until it starts to turn green, at which point it is no longer effective and should be discarded safely. If a cell is cleaned too long or with too strong a solution, its walls gradually lose their landmarks, the inner surfaces can no longer be found visually, and the cell becomes unusable. It is important to keep the cell filled with deionized water when it is not in use. Before the next run, the cell is rinsed with 1 mM Na$_{2.93}$ EDTA (diluted from the 100 mM stock described above), then water, then is filled and emptied twice (once from each side) with the electrolyte to be used in the electrophoresis experiment. No detergent is ever used, because it is virtually impossible to remove entirely, and it can create havoc with bubble formation and liposome contamination.

The round-bottom glass flask in which the lipid films are made requires special attention, because the dried lipids contact this surface intimately. The cleaning is similar to that of the cell, including the chromic/sulfuric acid treatment. Other glassware is always rinsed with EDTA, but never detergent.

After a run, the current electrodes are rinsed in water, acetone, chloroform–methanol (2:1, v/v), then acetone, 1 mM EDTA, and water. They are dried in air before their next use.

Measurements: Electrophoretic Mobility

When the microscope is focused at a stationary layer, there should be at least 20 liposomes in sharp focus. Before the electric field is switched on, the liposomes are checked for drift. Convective drift usually results from air in the cell. The problem is solved by removing the electrodes, topping up the liposome–electrolyte solution, and reseating the electrodes. Small drifts are averaged out by the averaging procedure described later. A "run" consists of 10 readings at the front stationary layer, with alternate reversals of electrode polarity, then 10 similarly at the rear stationary layer, for a total of 20 readings. Liposomes are selected randomly, one at a time, including all sizes from all parts of the visual field. The time for the selected liposome's image to traverse horizontally a given number of squares on the eyepiece grid is recorded with a digital timer to $\sim 0.01\,$s, and at the same time the electrometer voltage is recorded to 0.01 V. The mobility is $v/E = $ (distance/time)/(volts/ℓ), where ℓ is the effective distance between the voltage electrodes, that is, the calibrated cell constant. The electrode reversals allow convective drift and electrode polarization effects to be averaged out (but they do not remove the effects of electroosmotic convection, as mentioned previously). However, such effects are minimal, especially with the four-probe cell. The mobilities at the two stationary layers are averaged separately and compared. If they differ by more than 15%, the run is discarded. Unequal readings at the two stationary layers mean that the hydrodynamics were probably perturbed by dirt or bubbles on the walls, or that the stationary layers were not found correctly. Errors due to aberrant stationary layers do not generally cancel each other. In any case, the data are unreliable. Comparison of mobilities at the two stationary layers is an important element of quality control. If the readings agree to within 15%, they are averaged.

It should be noted that the quantities being averaged in the previously-described protocol are the mobilities, which are proportional to $1/tV$, where t is the measured transit time (in seconds) and V is the measured voltage (in volts). The average $\langle 1/tV \rangle$ is calculated by taking the reciprocal of tV for each measurement, and then averaging these reciprocals. (This is not the same as first averaging tV, then taking its reciprocal, i.e., $1/\langle tV \rangle$, which would give an incorrect result.) The average mobility is $\langle 1/tV \rangle \times \ell \times$ grid spacing $\times$ number of grid squares counted. The Smoluchowski zeta potential (in millivolts) is $12.80 \times$ mobility (in μm·cm/V·s); see Eq. (3). It is convenient to calculate the constant $B \equiv 12.80 \times \ell \times$ grid spacing, with ℓ in centimeters and the grid spacing in micrometers. At each stationary layer the average Smoluchowski zeta potential is then $B \times$ number of grid squares counted $\times \langle 1/tV \rangle$.

The number of eyepiece grid squares counted during a measurement is the experimenter's choice. On average, it takes a liposome twice as long to traverse twice as many squares, so its mobility is independent of the number of squares counted. However, the statistical accuracy increases with the elapsed time, which should be at least several seconds. For fast-moving liposomes, 10 grid squares are counted, whereas for very slow-moving ones, 1 square is enough.

For each sample investigated, at least three runs are performed, the cell being emptied, cleaned, and refilled for each new run. One or more additional sets of runs are performed at a later time with different batches of lipids and buffers, and all results are averaged.

As controls, we find that bovine PS multilamellar liposomes in 100 mM NaCl, 10 mM MOPS, 1 mM EDTA, pH 7.2 have a zeta potential of -58.2 ± 0.2 mV, and that multilamellar palmitoyloleoylphosphatidylglycerol–palmitoyloleoylphosphatidylcholine (POPG–POPC) 1:9 (mol/mol) liposomes in 10 mM NaCl, 1 mM MOPS, 0.1 mM EDTA, pH 7.2 have a zeta potential of -61.2 ± 0.3 mV, in agreement with the results of previous workers.[2,8]

Other Devices

The main disadvantage of the Rank Brothers machine in the flat-cell configuration is that liposomes of size smaller than ~1 μm cannot be visualized. For smaller particles such as unilamellar liposomes, other devices must be used. For these devices, many of the previously-described principles apply, the main difference being the method used to detect the particle velocities. In this section we mention some features, advantages, and disadvantages of the Coulter Delsa (Beckman Coulter, Hialeah, FL) and the ZetaPALS (Brookhaven Instruments, Holtsville, NY). Another system, with which we have not had experience, is the ZetaSizer (Malvern, Southborough, MA).

The Delsa uses a technique known as laser-Doppler interferometry. Four laser beams are scattered off the particle suspension at different angles. The particle velocities Doppler-shift the frequency of the scattered light, which is then analyzed to infer the velocity distribution.[10] Whereas the Rank Brothers machine tracks individual liposomes, the Delsa obtains the velocity spectrum of the entire suspension at once. The use of four angles permits detection of a wide range of particle sizes and helps eliminate artifacts. Both multilamellar and extruded unilamellar liposomes can be measured. The cell volume is less than 1 ml. The Delsa can also be used to measure particle sizes, but not as well as the Coulter N4 Plus submicron particle sizer. Some problems are as follows: zeta potentials cannot be measured near zero ($\lesssim$ 8 mV), an artifactual stationary peak centered at zero velocity obscures data for weakly scattering samples, loading the cell requires skill to avoid bubbles (solutions must be degassed), the stationary layers must be found (similar to the Rank Brothers machine), the cell is oriented horizontally and thus is susceptible to sedimentation errors, the software is cumbersome (has not been updated since 1995), and the cell is expensive. The cell is integrated with either silver or gold electrodes. The gold cell requires less frequent cleaning than the silver cell and is recommended for liposome work. Rinsing the cell between samples is simple, but thorough cleaning (needed about once per month with the gold cell) requires complete disassembly of the cell. We find the Delsa results to be reproducible, and agreement with the Rank Brothers machine for multilamellar liposomes is good.

The Brookhaven Instruments ZetaPALS uses a different technique called phase analysis light scattering, in which the phase variation of scattered light in the time domain is analyzed to yield the average velocity of the particle suspension.[11,12] It uses an Uzgiris cell employing two flat electrodes positioned far from the walls of a disposable plastic cuvette. Thus there are no stationary layers or sedimentation problems, although the electrodes do require routine cleaning. Solutions must be degassed and filtered to remove dust. The cell volume is ~1.5 ml. This instrument can measure close to zero velocity. However, it does not yield information about the velocity distribution—only the mean velocity is determined. Other advantages are as follows: it works well for small particles,

[10] R. Xu, "Particle Characterization: Light Scattering Methods." Kluwer Academic, Dordrecht, The Netherlands, 2000.

[11] J. F. Miller, K. Schätzel, and B. Vincent, *J. Colloid Interface Sci.* **143**, 532 (1991).

[12] W. W. Tscharnuter, F. McNeil-Watson, and D. Fairhurst, *ACS Symp. Series* **693**, 327 (1998).

requires only small electric fields, measures particle sizes, has high through-put, the correlator and software are excellent, and the company was founded and is managed by scientists. Both the Delsa and ZetaPALS employ two-probe cells, which lack the advantages of the Rank Brothers four-probe cell.

Interpretation of Results

Relation between Electrophoretic Mobility and Liposome Charge

Here we give the formulas needed to calculate the electrostatic surface potential and net charge of a liposome once its electrophoretic velocity has been measured.

For experimentally realizable electric fields, the electrophoretic velocity (v_e) is linear in the applied field strength (E). Therefore, as men-tioned, the particle motion is expressed in terms of velocity per unit field, known as the electrophoretic mobility ($\mu_e \equiv v_e/E$). There is no general ana-lytic formula for the charge of a spherical hydrodynamic particle in terms of its electrophoretic mobility. Exact formulas exist for limiting cases, and excellent approximate expressions exist for a wide range of conditions. Otherwise, numerical solutions may be used. (The general problem for spheres has been solved numerically by O'Brien and White.[13] A Fortran program for performing this computation is available.[14])

The procedure for determining the charge is as follows.

1. From the experimentally measured mobility, the electrostatic poten-tial is calculated at the surface of the hydrodynamic particle. This potential is known as the zeta potential (ζ). The equation used is the Smoluchowski equation or other approximate formulas depending on the value of κa (see p. 170–171). Alternatively, a numerical calculation can be done with the O'Brien and White program.

2. To proceed from the hydrodynamic particle to the "physical" particle requires knowledge of the location of the shear surface relative to the particle's physical surface. If this distance is known (or assumed), the electrostatic potential at the particle surface can be calculated from the ζ potential. This is the electrostatic surface potential (ψ_0). The relation used is the Poisson–Boltzmann electrostatic potential profile, in either planar or spherical form.

[13] R. W. O'Brien and L. R. White, *J. Chem. Soc. Faraday II* **74,** 1607 (1978).
[14] Requests may be made to Prof. D. Y. C. Chan, Department of Mathematics and Statistics, University of Melbourne, Parkville, Victoria 3010, Australia.

3. From the surface potential, the net charge (or surface charge density σ_0) of the particle may be calculated by the Gouy (or Grahame) equation, in either planar or spherical form.

Calculation of Zeta Potential

The correct analytic formula with which to calculate the zeta potential from the mobility depends on the quantity κa, where κ is the Debye–Hückel constant of the electrolyte and a is the radius of the particle (assumed spherical). Because $\kappa \equiv 1/\lambda_D$, where λ_D is the Debye screening length of the electrolyte, it follows that κa is the ratio of the particle radius to the Debye length.

The formula for κ is: $\kappa = (2e^2 I/\varepsilon kT)^{1/2}$, where e is the electron charge, I is the ionic strength of the electrolyte, ε is the dielectric permittivity of the electrolyte (which $= \varepsilon_r\varepsilon_0$, where ε_r is the dimensionless dielectric constant and ε_0 is the permittivity of free space), k is Boltzmann's constant, and T is the absolute temperature. For an aqueous electrolyte at $25°$, $\kappa = 0.3286$ $I^{1/2}$, where κ is in units of $Å^{-1}$ and I is in units of molarity. Equivalently, $\lambda_D = 3.043 I^{-1/2}$, with λ_D in angstroms. As a rule of thumb, at $25°$ the Debye length is approximately (within 4%) 100 Å at 1 mM ionic strength, 30 Å at 10 mM, 10 Å at 100 mM, and 3 Å at 1 M.

When $\kappa a \gg 1$ (i.e., $\lambda_D \ll a$) the screening layer is said to be "thin" compared to the size of the particle. In practical terms, for liposomes suspended in 10 mM electrolyte (often used for these measurements), this criterion is met when the liposome diameter $\gg 60$ Å, which is nearly always the case. For example, in 10 mM electrolyte, 100-nm liposomes have $\kappa a \simeq 16$, which is in the "thin layer" regime.

The limit of very large κa is known as the "Smoluchowski limit." When $\kappa a \rightarrow \infty$, the screening layer is infinitesimally thin. This means, because κ is finite, that a is infinitely large, that is, the particle surface is flat. In this limit the relation between mobility and zeta potential is given exactly by the Smoluchowski equation:

$$\zeta = (\eta/\varepsilon)\mu_e \tag{3a}$$

where η and ε are the solvent viscosity and dielectric permittivity, respectively. The units of mobility are v_e/E, where v_e is the electrophoretic velocity, usually in units of micrometers per second; and E is in units of volts per centimeter. For water at $25°$, the Smoluchowski equation, in useful units, is

$$\zeta = 12.80\mu_e \tag{3b}$$

where ζ is in millivolts and μ_e is in units of μm·cm/V·s. For finite κa, Eq. (3) underestimates the true zeta potential. The accuracy of Eq. (3) depends on the magnitudes of κa and ζ. For $\kappa a > 500$, known as the Smoluchowski regime, Eq. (3) may be used with $> 95\%$ accuracy for all but very large zeta potentials.[1] For a 10 mM electrolyte, $\kappa a > 500$ corresponds to a liposome diameter $> 3 \mu$m, which is the approximate size of a multilamellar liposome. For smaller liposomes, ζ becomes size dependent, and Eq. (3) may lead to significant errors unless ζ is very small. The extent to which Eq. (3) underestimates the zeta potential for particular values of κa and ζ may be estimated from the numerical plots of O'Brien and White[13] (their Fig. 4), which are reproduced in Hunter[1] (his Fig. 3.19). For $\kappa a < 500$ the approximate analytical formula of O'Brien and Hunter[15] [their Eq. (4.20), see also Hunter,[1] Eq. (3.7.14)] may be used down to $\kappa a \approx 30$ with no more than a few percent error. The more complicated semiempirical formula of Ohshima *et al.*[16] [their Eq. (77)] may be used down to $\kappa a \approx 10$ with less than 1% error.

Thus, to select the correct formula for calculating the zeta potential from the measured mobility, κa must be determined. κ is known from the electrolyte ionic strength, but a requires an independent size measurement. We use the Coulter N4 Plus submicron particle sizer (Beckman Coulter), which measures particle sizes ranging from 3 nm to 3 μm. For micrometer-sized liposomes, a direct visual estimate can also be made via the calibrated eyepiece grid on the Rank Brothers machine. For $\kappa a > 500$, the relation of ζ to μ_e is virtually independent of size (and shape), so accurate size measurements are not necessary.

The regime of "thick" screening layers ($\kappa a \ll 1$) is generally not relevant for liposomes and is not discussed here. Hunter[1] may be consulted for details.

Calculation of Surface Potential

The electrostatic surface potential ψ_0 can be calculated from ζ if the distance of the liposome surface from the shear surface is known (or assumed). The "liposome surface" generally is taken to mean the charge layer defined by the phospholipid phosphate groups, although this convention is not universal. The Poisson–Boltzmann electrostatic potential profile is used to find ψ_0 from ζ. We consider four cases, restricting our discussion to 1:1 (uni-univalent) electrolytes. Generalizations for more complex electrolytes may be found in the references.

[15] R. W. O'Brien and R. J. Hunter, *Can. J. Chem.* **59,** 1878 (1981).
[16] H. Ohshima, T. W. Healy, and L. R. White, *J. Chem. Soc. Faraday Trans. 2* **79,** 1613 (1983).

Planar Case, Small Potentials. In the Smoluchowski regime ($\kappa a > 500$) the liposome surface can be approximated as planar. If the potentials are small ($e\psi_0/2kT \lesssim 1$, or $\psi_0 \lesssim 50\,\text{mV}$ at $25°$), the potential profile in the electrolyte is given by the Debye–Hückel expression:

$$\psi(x) = \psi_0 \exp(-\kappa x) \tag{4}$$

where x is the perpendicular distance from the planar surface, $\psi_0 \equiv \psi(0)$, and κ is the Debye–Hückel constant. Because $\zeta \equiv \psi(\Delta x)$, where Δx is the distance of the shear surface from the liposome surface, Eq. (4) yields

$$\zeta = \psi_0 \exp(-\kappa \Delta x) \tag{5}$$

whence

$$\psi_0 = \zeta \exp(\kappa \Delta x) \tag{6}$$

Planar Case, Arbitrary Potentials. If the potentials are too large to justify use of the Debye–Hückel approximation, Eq. (4) is replaced by[3,8]

$$\psi(x) = \frac{2kT}{e} \ln\left[\frac{1 + \alpha \exp(-\kappa x)}{1 - \alpha \exp(-\kappa x)}\right] \quad \text{where} \quad \alpha \equiv \frac{\exp(e\psi_0/2kT) - 1}{\exp(e\psi_0/2kT) + 1} \tag{7}$$

Equation (6) becomes

$$\psi_0 = \frac{2kT}{e} \ln\left[\frac{1 + \alpha}{1 - \alpha}\right] \quad \text{where} \quad \alpha = \left(\frac{\exp(e\zeta/2kT) - 1}{\exp(e\zeta/2kT) + 1}\right) \exp(\kappa \Delta x) \tag{8}$$

At $25°$, $2kT/e = 51.39\,\text{mV}$. The error caused by using Eq. (6) instead of Eq. (8) depends on the magnitudes of ζ and $\kappa \Delta x$. For $\zeta \approx 50\,\text{mV}$ in $10\,\text{m}M$ NaCl (using $\Delta x = 4$ Å), Eq. (6) underestimates ψ_0 by $\sim 2.5\%$. Equations (7) and (8) reduce to Eqs. (4) and (6) when ψ_0 and ζ are small ($\ll 50\,\text{mV}$ at $25°$).

Spherical Case, Small Potentials. When the screening layer is not thin enough (i.e., κa not large enough) to justify use of the planar approximation, the spherical form of the electrostatic potential profile must be used. For small potentials, the Debye–Hückel expression in spherical coordinates is[1]

$$\psi(r) = \psi_0\, a\, \frac{e^{-\kappa(r-a)}}{r} \tag{9}$$

where r is the radial distance from the center of the sphere, a is the sphere radius, $\psi_0 \equiv \psi(a)$, and $r \geq a$. If Δr is the distance of the shear surface from the sphere surface, then $\zeta \equiv \psi(a + \Delta r)$. Therefore,

$$\zeta = \psi_0\, a\, \frac{e^{-\kappa \Delta r}}{a + \Delta r} \tag{10}$$

whence

$$\psi_0 = \zeta\left(1 + \frac{\Delta r}{a}\right)\exp(\kappa\Delta r) \tag{11}$$

When $\Delta r \ll a$, the geometry is nearly planar, and Eq. (11) reduces to Eq. (6). The error caused by using Eq. (6) instead of Eq. (11) depends on $\Delta r/a$. For 100-nm liposomes in 10 mM NaCl, with $\Delta r = 4$ Å and $a = 500$ Å, Eq. (6) underestimates ψ_0 by less than 1%. For liposomes with surface-grafted coats, however, $\Delta r/a$ can be much larger,[3] in which case Eq. (11) should be used in place of Eq. (6).

Spherical Case, Arbitrary Potentials. When the potentials are too large for the Debye–Hückel approximation ($\gtrsim 50$ mV at 25°), the spherical analog of Eq. (7) must be used. Unfortunately, there is no general analytic solution to the nonlinearized Poisson–Boltzmann equation in spherical coordinates. However, an accurate approximate analytic solution (differing from the numerical solution by $<1\%$) has been reported for $\psi(r)$ up to at least $5kT/e \approx 125$ mV for arbitrary κa.[17] This expression may be used to find ψ_0 in terms of ζ and Δr in the same way Eqs. (6), (8), and (11) are derived above.

Calculation of Particle Charge or Surface Charge Density

Once the surface potential ψ_0 is known, the charge of the physical particle may be determined. From ψ_0 and the electrostatic potential profile $\psi(x)$ or $\psi(r)$, the volume charge-density profile in the diffuse screening layer is obtained from the Poisson equation, and the total charge of the screening layer is found by integration. Global electroneutrality then requires that the charge of the particle equals minus the total charge in the screening layer. Again we discuss four cases.

Planar Case, Small Potentials. For the planar Debye–Hückel potential [Eq. (4)], the result is

$$\sigma_0 = \varepsilon\kappa\psi_0 \tag{12a}$$

where σ_0 is the surface charge density of the particle (usually in elementary charges/Å^2 or μC/cm^2). At 25° in useful units,

$$\sigma_0 = (4.340 \times 10^{-4})\kappa\psi_0 = (1.426 \times 10^{-4})\,C^{1/2}\psi_0 \tag{12b}$$

where σ_0 is in units of $e/$Å^2 (i.e., the number of |electron charges| per Å^2), ψ_0 is in millivolts, κ is in Å^{-1}, and C is the uni-univalent electrolyte concentration in molar units. For liposomes, surface-charge density units $e/$Å^2 are

[17] R. Tuinier, *J. Colloid Interface Sci.* **258**, 45 (2003).

useful as they can be related to molecular quantities via the area per lipid. For example, if the area per lipid on the outer leaflet of a liposome is ~70 $\mathring{A}^2$ and $\sigma_0 = -1.43 \times 10^{-2}$ charges/$\mathring{A}^2$, then on the average each phospholipid has one net negative charge. It is useful to note that σ_0^{-1} is the membrane area ($\mathring{A}^2$) per charge. To convert charge density to units of $\mu C/cm^2$, σ_0 is multiplied by 1.602×10^3 $(\mu C/cm^2)/(e/\mathring{A}^2)$.

Planar Case, Arbitrary Potentials. For the planar Poisson–Boltzmann potential [Eq. (7)], the result for a uni-univalent electrolyte is

$$\sigma_0 = \frac{2kT}{e} \varepsilon\kappa \sinh \frac{e\psi_0}{2kT} \tag{13a}$$

At 25°,

$$\sigma_0 = (2.230 \times 10^{-2})\kappa \sinh \frac{\psi_0}{51.39} = (7.330 \times 10^{-3})C^{1/2} \sinh \frac{\psi_0}{51.39} \tag{13b}$$

where σ_0 is in units of $e/\mathring{A}^2$, ψ_0 is in millivolts, κ is in $\mathring{A}^{-1}$, and C is the monovalent salt concentration in units of molarity. Equation (13) is the Gouy equation. (A more general expression for mixed electrolytes is called the Grahame equation.[1,18,19]) Because $\sinh(x) \approx x$ for small x, Eq. (13) reduces to Eq. (12) when ψ_0 is small. The error caused by the use of Eq. (12) instead of Eq. (13) depends on ψ_0. For $\psi_0 \approx 50$ mV, Eq. (12) underestimates σ_0 by ~14%. Equation (12) approximates Eq. (13) to within 1% when $\psi_0 \lesssim 12$ mV. Thus, Eq. (13) should be used for all but the smallest potentials.

Spherical Case, Small Potentials. For the spherical Debye–Hückel potential [Eq. (9)], the particle charge in terms of ψ_0 is given by[1] $Q = 4\pi\varepsilon a(1 + \kappa a) \psi_0$, or

$$\sigma_0 = \varepsilon\kappa \left(1 + \frac{1}{\kappa a}\right)\psi_0 \tag{14a}$$

where $\sigma_0 \equiv Q/4\pi a^2$. At 25° in useful units,

$$\sigma_0 = (4.340 \times 10^{-4})\kappa \left(1 + \frac{1}{\kappa a}\right)\psi_0 = 10^{-4}\left(1.426\, C^{1/2} + \frac{4.340}{a}\right)\psi_0 \tag{14b}$$

with σ_0 in charges/$\mathring{A}^2$, ψ_0 in millivolts, κ in $\mathring{A}^{-1}$, a in angstroms, and C in units of molarity. When $\kappa a \gg 1$, Eq. (14) reduces to Eq. (12). The error caused by the use of Eq. (12) instead of Eq. (14) depends on κa. For 100-nm liposomes in 10 mM NaCl, where $\kappa a \approx 16$, Eq. (12) underestimates σ_0 by ~6%.

[18] S. McLaughlin, *Curr. Top. Membr. Transp.* **9**, 71 (1977).
[19] P. C. Hiemenz, "Principles of Colloid and Surface Chemistry," 2nd Ed. Marcel Dekker, New York, 1986.

Spherical Case, Arbitrary Potentials. There is no exact analytical equation relating σ_0 to ψ_0 for a spherical geometry when $\psi_0 \gtrsim 50\,\text{mV}$ at 25°. This situation has been discussed by Hunter,[1] who recommends the semiempirical formula of Loeb *et al.*,[20] which approximates their numerical solution to better than 1% for ψ_0 as high as $150\,\text{mV}$ and κa as low as 2:

$$\sigma_0 = \frac{2kT}{e}\,\varepsilon\kappa\left(\sinh\frac{e\psi_0}{2kT} + \frac{2}{\kappa a}\tanh\frac{e\psi_0}{4kT}\right) \tag{15a}$$

At 25°,

$$\begin{aligned}
\sigma_0 &= (2.230 \times 10^{-2})\kappa\left(\sinh\frac{\psi_0}{51.39} + \frac{2}{\kappa a}\tanh\frac{\psi_0}{102.8}\right) \\
&= (7.330 \times 10^{-3})\left(C^{1/2}\sinh\frac{\psi_0}{51.39} + \frac{6.086}{a}\tanh\frac{\psi_0}{102.8}\right)
\end{aligned} \tag{15b}$$

with σ_0 in charges/Å^2, ψ_0 in millivolts, κ in Å^{-1}, a in angstroms, and C in units of molarity. Because $\sinh(x) \approx \tanh(x) \approx x$ for small x, Eq. (15) reduces to Eq. (14) when ψ_0 is small, and it reduces to Eq. (13) when κa is large. For a 100-nm liposome in $10\,\text{m}M$ NaCl with $\psi_0 \approx 50\,\text{mV}$, Eq. (13) underestimates σ_0 as calculated from Eq. (15) by $\sim 5\%$, and Eq. (14) underestimates it by $\sim 14\%$. An improved, but more complicated, version of Eq. (15) has been given by Ohshima *et al.*[21]

To calculate the surface charge density σ_h of the hydrodynamic particle, Eq. (12), (13), (14), or (15) is used with ζ substituted for ψ_0, and $a + \Delta r$ substituted for a. In this case Eqs. (4)–(11) are not needed. The total charge Q_h of the hydrodynamic particle is $Q_h = 4\pi(a + \Delta r)^2\,\sigma_h$.

Ion Binding

The quantity σ_0 calculated in Eqs. (12)–(15) is the *net* surface charge density of the particle, with no specification as to the microscopic origin of the charge. For the previous example of PS multilamellar liposomes in $\sim 111\,\text{m}M$ NaCl with $\zeta = -58.2\,\text{mV}$, κa is ~ 2000. Taking $\Delta x = 2\,\text{Å}$, Eqs. (8) and (13) yield $\sigma_0 = -5.26 \times 10^{-3}\,e/\text{Å}^2$, or $\sigma_0^{-1} = 190\,\text{Å}^2$ per negative charge. Compared with the PS area per lipid $\sim 70\,\text{Å}^2$, it is apparent that there is much less than one negative charge per PS headgroup on the membrane. This discrepancy has been attributed to ion binding.[8] By use of the Stern adsorption isotherm,[18,22] it can be shown that a Na · PS

[20] A. L. Loeb, J. T. G. Overbeek, and P. H. Wiersema, "The Electrical Double Layer around a Spherical Colloid Particle." MIT Press, Cambridge, MA, 1961.

[21] H. Ohshima, T. W. Healy, and L. R. White, *J. Colloid Interface Sci.* **90,** 17 (1982).

[22] J. A. Cohen and M. Cohen, *Biophys. J.* **36,** 623 (1981).

intrinsic binding constant of $0.75\ M^{-1}$ is consistent with the observed ζ and the above-calculated value of σ_0. Proton binding can be treated similarly.

Surface Structure and Hydrodynamic Drag

Inclusion of poly(ethylene glycol)-linked phospholipids in the liposome-forming mixture produces liposomes with "fuzzy" surfaces.[3] Colloidal particles with such surfaces are known as "soft" particles.[23] The electrophoretic mobilities of soft liposomes are decreased by hydrodynamic drag caused by friction between the fuzzy coat and the water. Thus a soft liposome bearing neutral polymers has a lower mobility than a hard liposome of the same size with the same surface charge density. One interpretation of such effects is that the effective shear surface moves outward by an amount related to the shape, thickness, and frictional properties of the polymer layer. Thus electrophoretic mobility can be used to infer properties of the surface polymer coat.[3]

Conclusions

Particle electrophoresis is a useful technique for measuring the net charge or surface charge density of liposomes. It also provides valuable information about liposome surface structure. Several commercial devices are available for measuring liposome electrophoretic mobilities, each having its own advantages and disadvantages. In all cases, the experiments must be done with well-calibrated instruments, well-controlled solutions, and precisely formulated lipid mixtures. In each case, care must be taken to use the appropriate equations to calculate the charge from the mobility.

It is important to appreciate that charge is not the sole determinant of electrophoretic mobility. For $\kappa a < 500$, mobilities are size and shape dependent. For variously sized liposomes having the same charge density, smaller liposomes move more slowly than larger ones. For liposomes of identical charge and size, those with rough or "soft" coats move more slowly than those that lack such structures. Proper consideration of these effects is necessary for accurate interpretation of electrophoretic mobility measurements.

Acknowledgments

The author thanks Dr. Valentina Khorosheva for expert assistance, Professor Wayne Saslow for helpful discussions, and Mr. Neal Johnson and Mr. Arnold Eilers for the graphics.

[23] H. Ohshima, *Adv. Colloid Interface Sci.* **62**, 189 (1995).

[12]   Viscometric Determination of Axial Ratio of Ellipsoidal DNA–Lipid Complex

By Sadao Hirota

Introduction

A capillary viscometer gives accurate viscosity and is used to determine the molecular weight and shape parameters of polymers, molecular interactions and structure formation in solutions of polymers, as well as more general specifications of liquids. It is an excellent tool for quality control and provides a method for testing the specifications of liposomes and DNA–lipid complex delivery systems. However, it has seldom been used to characterize liposomes. The reasons are that it requires a large amount of sample and that it necessitates long, strenuous attention of the researcher during flow time measurements.

Here, an automation of viscosity measurements with a minicapillary viscometer is presented and its application to characterization of liposomes and DNA complexes is described.

Viscosity of Small Unilamellar Vesicle Suspension

Small unilamellar vesicles (SUVs) are known to be spherical particles. They are nonattracting in dilute aqueous suspension. For suspensions of spherical nonattracting particles, the Einstein equation[1] is given by

$$\eta_{sp} = (\eta/\eta_0) - 1 = 2.5\phi \quad \text{or} \quad \eta_{sp}/\phi = \eta_{red} = 2.5 \tag{1}$$

where η_{sp} is the specific viscosity, η is the observed viscosity of the suspension, η_0 is the viscosity of the suspending medium, η/η_0 is the relative viscocity, ϕ is the volume fraction of the suspended particles, and η_{red} is the nondimensional reduced viscosity. Equation (1) has been confirmed to hold experimentally by Jeffrey and by Cheng and Schachman.[2]

$$\phi = \text{Volume of particles}/\text{Volume of suspension} \tag{2}$$

[1] A. Einstein, *Ann. Phys.* **19,** 289 (1906).
[2] B. B. Jeffrey, *Proc. R. Soc.* (*Lond.*) *A* **102,** 163 (1923); P. Y. Cheng and H. K. Schachman, *J. Polym. Sci.* **16,** 19 (1955).

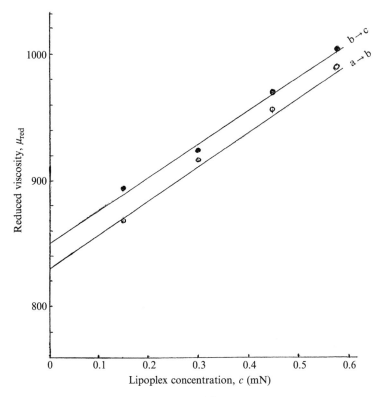

FIG. 1. Determination of intrinsic viscosity, $[\mu]$, of DNA–lipid complex. *Sample:* Plant DNA, sonicated to 0.5 to 2 kbp, is complexed with distearyldimethylammonium chloride (DDAC, FW 575.5) at a molar ratio of 1:4, and dissolved in 10 mM Tris-HCl buffer (pH 8.0). The complex has an FW of 2490 and a net density of 1.001. The final DNA concentration in the stock solution is 0.57 mN of the base unit. The stock solution is diluted to 0.149, 0.297, and 0.446 mN. *Measurements:* Viscosity is measured by automatic minicapillary viscometer at $40.00 \pm 0.01°$. *Data:* t_{10}, t_{20}: Flow time (seconds of 10 mM Tris-HCl buffer; t_1, t_2: flow time of lipoplex suspension, $\eta_{rel} = t_1/t_{10}, t_2/t_{20}$ through the same reservoir, $\eta_{sp} = \eta_{rel} - 1$, $\mu_{red} = \eta_{sp}/c$.

Although the viscosity of a suspension of attracting particles gives higher η_{sp} than is expected from Eq. (1), if ϕ is extrapolated to zero, it is confirmed that Eq. (1) is still valid:

$$\lim_{\phi \to 0} \eta_{red} = [\eta] = 2.5 \qquad (3)$$

where $[\eta]$ is the nondimensional intrinsic viscosity (Fig. 1).

Determination of Average Volume Fraction for Small Unilamellar Vesicles

Although Eq. (1) holds irrespective of particle size when the composition of particle material is homogeneous, in the case of SUVs ϕ is not a fixed value but depends on the particle size distribution, because the volume fraction of the inner aqueous phase of SUVs varies with the size. Note that the composition of SUVs is not homogeneous: the amount of water varies.

Here, let us assume that, if we use an average value of volume fraction, Eq. (3) still holds:

$$\lim_{\phi_{av} \to 0} \eta_{sp}/\phi_{av} = [\eta]_{av} = 2.5 \tag{4}$$

where ϕ_{av} is the volume fraction of SUVs that is to be determined. Intrinsic viscosity $[\mu]$ is defined by using molality, c (mol/kg), as

$$[\mu] = \lim_{c \to 0} \eta_{sp}/c \tag{5a}$$

as $[\mu]$ is proportional to $[\eta]_{av}$:

$$[\mu] = k[\eta]_{av} = 2.5k \tag{5b}$$

$[\mu]$ can be determined experimentally. Reduced viscosity, $\mu = \eta_{sp}/c$, is determined at several concentrations, c. $\mu = \eta_{sp}/c$ values are plotted against c. The plot is similar to Fig. 1. Then, μ is extrapolated to zero concentration to give intrinsic viscosity, $[\mu]$. The value of $k = [\mu]/2.5$ is then found and ϕ_{av} is given by

$$\phi_{av} = kc \tag{5c}$$

The volume fraction of the inner aqueous phase, ϕ_i, is given by

$$\phi_i = \phi_{av} - \phi_{net} \tag{6}$$

where ϕ_{net} is the net volume fraction of lipid material calculated by

$$\phi_{net} = C/\rho \tag{7}$$

where C is the weight fraction of the lipid or lipoplex material and ρ is the specific gravity of the lipid or lipoplex material. Here, ρ is estimated from chemical tables. Table I shows some examples of specific gravities of lipid materials. In most cases of lipoplex, $\rho = 1$ may be used. The volume fraction of the inner aqueous phase is usually determined by the encapsulation ratio of a fluorescent dye. For quality control in production, the dye cannot be used. Viscometric characterization of the inner aqueous phase is useful, especially in process control of SUV productions. This technique is also useful when bioactive ingredients are affected by the dye.

TABLE I
SPECIFIC GRAVITIES OF LIPID MATERIALS OF LIPOSOMES

	Specific gravity	d 25/25
Water		1.0001
Dimyristoylphosphatidylcholine (DMPC)	[1.032]	
Dipalmitoylphosphatidylcholine (DPPC)		1.0325
Disteatroylphosphatidylcholine (DSPC)	[1.035]	
Dioleoylphosphatidylcholine (DOPC)	[1.04]	
Dimyristoylphosphatidylglycerol (DMPG)	[1.04]	
Dipalmitoylphosphatidylglycerol (DPPG)	[1.04]	
Disteatroylphosphatidylglycerol (DSPG)	[1.04]	
Dioleoylphosphatidylglycerol (DOPG)	[1.04]	
Dimyristoylphosphatidylethanolamine (DMPE)	[1.04]	
Dipalmitoylphosphatidylethanolamine (DPPE)	[1.04]	
Disteatroylphosphatidylethanolamine (DSPE)	[1.04]	
Dioleoylphosphatidylethanolamine (DOPE)	[1.03]	
PEG 2000 dimyristoylphosphatidylethanolamine (PEG 2000 DMPE)	[1.02][a]	
PEG 2000 dipalmitoylphosphatidylethanolamine (PEG 2000 DPPE)	[1.02][a]	
PEG 2000 disteatroylphosphatidylethanolamine (PEG 2000 DSPE)	[1.02][a]	
PEG 2000 dioleoylphosphatidylethanolamine (PEG 2000 DOPE)	[1.02][a]	
Cholesterol (CH)		1.052
Myristic acid		0.8719
Stearic acid		0.9425
Oleic acid		0.9823
Cetylamine		0.8144
Stearylamine		0.8634
Oleylamine	[0.87]	
Cetyltrimethylammonium bromide (CTAB)	[0.86]	
Distearyldimetylammonium chloride (DSDMAC)	[0.87]	
Dioleoyltrimethylaminopropane chloride (DOTAP)	[0.87]	

[a] The true value is higher, but when it is on the SUV surface many water molecules are fixed in the PEG layer and 1.02 is considered practical.

Viscometry of Multilamellar Vesicle Suspension

Multilamellar vesicles (MLVs) are almost ellipsoidal, and Eq. (1) is no longer applicable to MLVs. Jeffrey[2] expanded the Einstein equation to ellipsoids and obtained the equation, $[\eta] = \nu$, and tried to relate ν to the axial ratio of ellipsoid, a/b. However, Jeffrey gave only ranges such as $2.174 < \nu < 2.819$ for $a/b = 2$ for prolate ellipsoid and $2.306 < \nu < 3.267$ for $b/a = 2$

for oblate ellipsoid. Simha[3] also extended the Einstein equation to ellipsoidal particles by replacing ν for 2.5 and related ν to the axial ratio, a/b, of the ellipsoid:

$$\lim_{\phi \to 0} \eta_{\text{red}} = [\eta] = \nu \tag{8}$$

Simha was successful in deriving equations in the range $a/b > 50$,

$$\nu = [J^2/15(\ln 2J - 1.5)] + [J^2/5(\ln 2J - 0.5) \\ + (14/15)] \quad \text{(for prolate ellipsoid)} \tag{9}$$

$$\nu = (16/15)(J/\tan^{-1}J) \quad \text{(for oblate ellipsoid)} \tag{10}$$

where J is a/b.

Equations (9) and (10) are not applicable to MLVs, because $a/b > 50$ is too large.

For studies on and quality control of MLVs DNA–lipid complexes, or protein, a definite value of a/b much smaller than 50 is required. However, Simha gave only the results of calculated numerical values for a/b in this region. Hirota[4] obtained equations relating nondimensional intrinsic viscosity, $[\eta]$, to axial ratios of ellipsoids.

In the range of $1 < a/b < 100$,

$$\nu = [\eta] = 0.057(a/b)^2 + 0.61a/b + 1.83 \quad \text{(for prolate ellipsoid)} \tag{11}$$

$$\nu = [\eta] = 0.001(a/b)^2 + 0.59a/b + 1.90 \quad \text{(for oblate ellipsoid)} \tag{12}$$

Thus, the axial ratio a/b of MLVs is easily determined experimentally by measuring the nondimensional intrinsic viscosity $[\eta]$. When $a/b = 1$, that is, the suspending MLV particles are perfect spheres, both Eqs. (11) and (12) give $[\eta] = 2.5$, reverting the Einstein Eq. (1). The relationship between the nondimensional intrinsic viscosity and the axial ratio of prolate and oblate ellipsoids is given in Table II and compared with Simha's results (Fig. 2).

According to the Sakurada equation,[5]

$$\ln[\eta] = \ln 2.5 + \alpha \ln M \tag{13}$$

where M is the nondimensional particle weight in daltons and α is the shape parameter. Here, α also characterizes the shape of MLVs. We can

---

[3] R. Simha, *J. Phys. Chem.* **44**, 25 (1940); R. Simha, *Science* **92**, 132 (1940).

[4] S. Hirota, *Biophysical Journal* **85**, 2 (2003).

[5] I. Sakurada, *Kagakusen-i kouenshuu* [in Japanese] **5**, 33 (1940); R. Houwink, *J. Prakt. Chem.* **157**, 15 (1940).

TABLE II
$[\eta]$ versus a/b Relationship: Comparison of Hirota, Jeffrey[a], and Simha Results

$[\eta]$	(a/b) prolate [Eq. (11)]	(a/b) oblate [Eq. (12)]
2.5	1	1
2.6	1.14	1.16
2.7	1.28	1.32
2.8	1.4	1.49
2.9	1.54	1.65
3	1.65	1.81
3.2	1.9	2.13
3.4	2.13	2.45
3.6	2.36	2.77
3.8	2.6	3.11
4	2.82	3.4
4.2	3.04	3.65
4.4	3.2	4
4.6	3.3	4.3
4.8	3.5	4.6
5	3.7	5
5.5	4.2	5.7
6	4.8	6.5
6.5	5.2	7.2
7	5.5	8

a/b	$[\eta]_{pr}$ [Eq. (11)]	$[\eta]_{ob}$ [Eq. (12)]
1	2.5	2.5
1.1	2.57	2.55
1.2	2.64	2.61
1.3	2.72	2.67
1.4	2.79	2.73
1.5	2.88	2.79
1.6	2.95	2.65
1.8	3.12	2.96
2	3.28	3.1
3	4.17	3.7
4	5.18	4.3
5	6.31	4.9
6	7.54	5.5
7	8.89	6.1
8	10.36	7
9	11.9	7.3
10	13.6	7.9
11	15.5	9
12	17.4	9.7
13	18.8	10.4

a/b	$[\eta]_{pr}$ by Jeffrey		$[\eta]_{ob}$ by Jeffrey		$[\eta]_{pr}$ by Simha[b]	$[\eta]_{ob}$ by Simha[b]
	Min	Max	Min	Max		
1	2.5		2.5		2.5	2.5
1.1						
1.2						
1.3						
1.4					2.63	2.62
1.5						
1.6						
1.8						
2	2.174	2.819	2.306	3.267	2.91	2.85
3					3.68	3.43
4					4.66	4.06
5					5.81	4.71
6					7.01	5.36
7						
8					10.1	6.7
9						
10	2.01	4.485	2.116	9.96	13.63	8.04
11						
12					17.76	9.39
13						

(continues)

a/b	$[\eta]_{pr}$	$[\eta]_{ob}$
7.5	5.9	8.7
8	6.3	9.6
8.5	6.7	10.3
9	7.1	11
9.5	7.4	11.7
10	7.7	12.4
11	8.3	13.8
12	9	15.3
13	9.6	16.8
14	10.2	18.2
15	10.7	19.5
20	13	26.8
25	15	33
30	17.3	40
40	21	50
50	24	65
100	35	134
150	45	190
200	57	240
250	61	290
300	67	325
350	76	360
400	80	400
500	90	465
1000	126	750

a/b	$[\eta]_{pr}$	$[\eta]_{ob}$
14	21.6	11.1
15	23.9	11.8
16	26.3	12.5
17	28.6	13.3
18	31.4	14
19	34.1	14.7
20	36.9	14.9
21	39.9	16.3
22	41.5	17
23	42.9	17.8
24	49.4	18.6
25	52.8	19.4
30	71.5	22.5
40	120	31.3
50	173	41.9
70	324	60
100	633	71
150	1376	110
200	2404	160
250	3718	210
300	5315	270
350	7199	330
400	9366	400
500	14,557	550
1000	57,612	1,590

a/b	$[\eta]_{pr}$	$[\eta]_{ob}$
14	24.8	11.42
15		
16		
17	38.6	14.8
18		
19		
20		
21		
22		
23		
24		
25	55.2	18.9
30	74.5	21.6
40	120.8	28.3
50	176.5	35
60	242	41.7
80	400.1	55.1
100	593	68.6
150	1222	102.3
200	2051	136.2
250		
300	4560	220
350	6026	260
400	7674	300
500	11,481	380
1000	40,271	760

[a] For prolate ellipsoid, $[\eta]_{pr} = 0.057\,(a/b)^2 + 0.61\,(a/b) + 1.83$ [Eq. (11)]; for oblate ellipsoid, $[\eta]_{ob} = 0.001\,(a/b)^2 + 0.59\,(a/b) + 1.9$ [Eq. (12)].

[b] From R. Simha, Science **92**, 132 (1940).

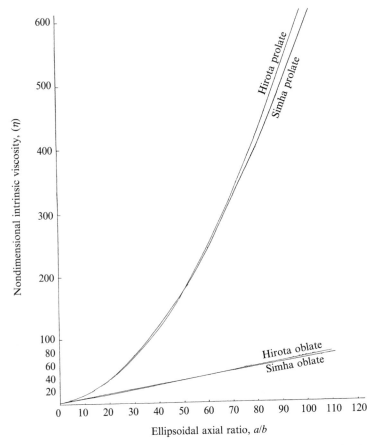

FIG. 2. Nondimensional intrinsic viscosity as a function of ellipsoidal axial ratio.

characterize shape both by the average value of the axial ratio, a/b, and by α of MLVs, as well as of DNA complexes, for research and quality control during large-scale production.

Viscometry of DNA–Lipid Complex and DNA–Polymer Complex

For DNA–lipid complexes (lipoplexes) and DNA–polymer complexes (polyplexes), the shape is one of the factors influencing transfection efficiency.[6–8] The shape can vary and has been described as beads on a string, fibril-like, tubules, hexagonal packs (beans), spaghetti and meatballs, map pins, and so on. Almost all these findings are by electron microscopy or atomic force microscopy, which reveals local information but does not give overall information or an average quantity required for quality control during large-scale production.

An average shape parameter of a DNA–lipid complex, expressed by the axial ratio of an ellipsoid, is given by viscometry using Eqs. (11) and (12). It is postulated that the complexation is complete and that there is no free DNA in the suspension. A value of the volume fraction is given by Eq. (7). For most cases, the complexes do not have an inner aqueous phase and the specific gravity of the complex particles can be determined experimentally when the complex has an average size of 100 nm or larger. An aliquot of the complex suspension is placed on the surface of an ultra-centrifuge tube containing a sucrose solution that has a concentration gradient: 10% at the bottom and 0% at the surface. After 10 min of centrifugation at $100\,g$ (10,000 rpm), the position of the turbid (opalescent) zone is observed. The sucrose concentration at this position gives the specific gravity of the complex particles by referring to Fig. 3. However, this method cannot be applied to liposomes with a large amount of cationic lipids, which have specific gravity less than 1.

Minicapillary Viscometer

A viscometer that uses less than 0.5 ml of liquid sample is necessary to study DNA–lipid complexes because the materials are usually expensive. To obtain flow times longer than 100 s for 0.5 ml to flow down a capillary 20 cm in length, the inner diameter of the capillary needs to be less than 0.5 mm. To obtain the intrinsic viscosity of a dilute aqueous solution of DNA, precision, in terms of coefficient of variation (CV), needs to be less than 0.3%. When the temperature of the thermostat is controlled to within ±0.01°, two significant figures of specific viscosity value ought to be obtained with a volume fraction less than 0.001 (about 1 mg/ml of the complex). The reproducibility of the measurements of flow time through an upper reservoir of 0.5 ml, expressed by standard deviation, is required to be less than 0.3 s (0.3%, in terms of CV) for a viscosity test of a DNA complex.

[6] P. L. Felgner, T. R. Gadek, M. Holm, R. Roman, H. W. Chan, M. Wenz, J. Northrop, G. M. Ringold, and M. Danielsen, *Proc. Natl. Acad. Sci. USA* **84**, 7413 (1987).
[7] H. R. Gershon, R. Ghirlando, S. B. Guttman, and A. Minsky, *Biochemistry* **32**, 7143 (1993).
[8] B. Sternberg, "Medical Applications of Liposomes" (D. D. Lasic, ed.), p. 395. Elsevier Science, New York, 1998.

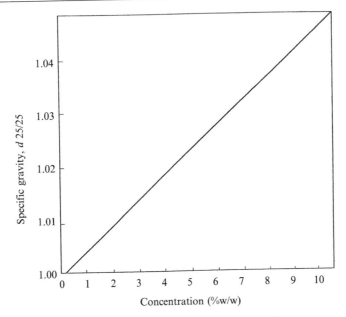

F<small>IG</small>. 3. Specific gravity of aqueous sucrose solution.

Problems with Minicapillary Viscometers

The reproducibility of the flow time, in terms of CV, is actually much larger than the above stated requirement, CV<0.3%. It is especially poor when the sample is an aqueous suspension of DNA–cationic lipid complex. The reason for the poor reproducibility is thought to be the formation of a water-repellent lipid monomolecular layer on the inner surface of the capillary. Even the adherence of an invisible air bubble has a highly significant effect on the flow time, because the flow time is inversely proportional to the fourth power of the capillary diameter. Trials to remove the water-repellant film with detergent or organic solvents have not been successful. Heating the capillary above 500° for 20 min in a furnace results in removal of the water-repellent film from the inner surface of the capillary (Fig. 4). A microcapillary viscometer is produced with heat-resistant glass, which remains stiff at 550° for 20 min. After heating the viscometer in the furnace, reproducibility of the flow time data is improved remarkably and meets requirements, even with measurements of suspensions containing cationic lipids. When the flow time increases and the CV exceeds 0.3%, the viscometer is rinsed and heated in the furnace. The original flow time data are then obtained again. It has been confirmed that repeated heating does not result in any changes in the inner diameter of the capillary.

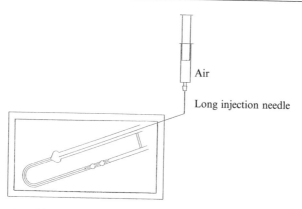

Fig. 4. Cleaning of inner capillary surface by heating the viscometer in a furnace. When a minicapillary viscometer is heated in a furnace at 500°, air is introduced into the lower reservoir through a long injection needle. The convection in the viscometer makes the removal of water-repellant films easy. Heating for 20 min is sufficient.

Automation of Viscosity Measurement

For classic viscosity measurements, the time for the sample liquid to flow down through a reservoir at the upper end of a capillary is measured by watching the passage of the meniscus through the orifices, using a stopwatch. An average of several measurements is compared with that of a standard liquid whose viscosity is well known. From the ratio of the sample flow time to that of the standard liquid, the relative viscosity of the sample is determined. Specific viscosity is obtained by subtracting unity from the relative viscosity. An absolute viscosity can be also determined by introducing the flow time, capillary diameter, capillary length, and pressure difference between both ends of the capillary into the Hagen–Poiseuille equation. Although the procedure is simple, it requires the long, strenuous attention of the observer and tiring repetitions.

We have developed an automated viscometer. The details of the micro-viscometer are shown in Fig. 5. When the meniscus passes through orifice a, b, or c, the laser beam sent from a laser source through an optical fiber produces a signal and reaches the sensor through an optical fiber. The signal is amplified at the relay switch. When the meniscus passes through orifice a, the signal starts a digital timer. When the meniscus passes through orifice b, the signal records the elapsed time, and when it passes through orifice c, the signal stops the digital timer, and the flow time from a to b, b to c, and a to c are stored in the memory of a digital calculator. When the meniscus passes orifice c, the signal opens an electromagnetic valve to a compressor. The compressed air pushes the sample liquid from the lower

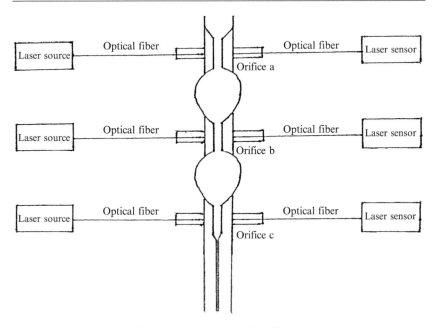

Fig. 5. Detailed side view of automatic capillary viscometer.

reservoir to the upper reservoir. When the meniscus passes upward through orifice a, the valve closes to the compressor and opens to the atmosphere. The second measurement of flow time then begins for the same sample to flow down from orifice a to c. The previously-described automatic measurements of flow time are repeated until the last measurement is made, turning off the relay switch. The recorded data in the memory of the digital calculator are averaged and displayed.

Example of Viscometry of Dilute Lipoplex Suspension by Automated Minicapillary Viscometer

Materials

Plant DNA, 10–50 kbp in 10 mM Tris-HCl buffer (pH 8.0), is sonicated in a bath sonicator, at 20 kHz, 200 W, at 10–40° for 20 min. This procedure is expected to reduce the size of DNA to 0.5–2 kbp, the approximate size of functional plasmid DNA. The concentration is 0.57 mN base units by optical density at 260 nm. The OD at 280 nm is also determined. The 260:280 OD ratio is 1.75, indicating that the DNA purity is more than 95%.

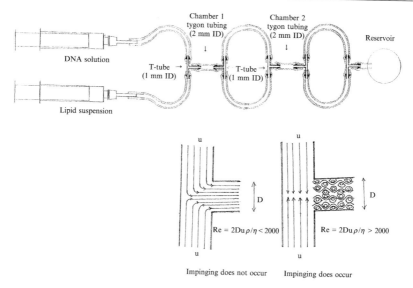

FIG. 6. Flow impinger.

Lipoplex Preparation by Flow Impinger

An excess of cationic lipid, 2.28 mN distearyldimethylammonium chloride (DDAC) in 10 mM Tris-HCl buffer (pH 8.0), and an equal volume of 0.57 mN DNA solution in 10 mM Tris-HCl buffer (pH 8.0) at a charge ratio of 1:4 (−/+), are mixed with a flow impinger (Fig. 6). This is an improvement of the Plexer reported previously.[9] A syringe is filled with DNA solution and another syringe is filled with lipid suspension. Both liquids are injected into the flow impinger at the same volume velocity of 1 ml/s. The two reaction liquids flow at the same velocity and impinge in T-tube 1 at an angle of 180°. A rapid stream forms turbulence in chamber 1 and is mixed uniformly. The mixture goes into the second T-tube. Thus, three successive impinging and mixing steps make the colloid chemical properties of the complex uniform, and the sample is released from the exit into a reservoir. When the volume velocity at each of the previously-described syringes, V, is 1 ml/s, the maximum shearing rate[10] in the arm of T-tubes, $\gamma = 4V/\pi r^3$, is 4000 (1/s) and the Reynolds number in the chamber with diameter D, $Re = DV \rho/\eta \pi r^2$, is 5000. If V is controlled in the range between 0.5–2.0 ml/s, satisfactory mixing and

[9] S. Hirota, C. Tros de Ilarduya, L. Barron, and F. Szoka, *Biotechniques* **27,** 286 (1999).
[10] R. B. Bird, *Transport Phenomena* **208,** 208 (1961).

homogenizing is attained without any danger of DNA strand breakage.[11] The transit time of each liquid through the flow impinger is 0.8 s at $V = 0.5$ ml/s and 0.2 s at $V = 2$ ml/s.

Viscosity Measurements

In Fig. 5, flow time t_{10} is the time required for the suspending medium to pass through reservoir 1, t_{20} is the time required for passage through reservoir 2, and t_{30} represents total time from reservoir a to c; t_1 is time required for passage of for sample through reservoir 1, t_2 for sample through reservoir 2, and t_s represents total time through reservoirs 1 and 2 are displayed from the recorder of the autominicapillary viscometer.

$$\eta_{rel} = t_1/t_{10} = t_2/t_{20} = t_3/t_{30} \cdots \quad \text{(Newtonian flow)} \quad (14)$$

$$\eta_{rel} = t_1/t_{10} < t_2/t_{20} \quad \begin{array}{l}\text{(plastic flow or orientation of} \\ \text{particles to the direction of} \\ \text{the flow)}\end{array} \quad (15)$$

$$\eta_{rel} = t_1/t_{10} > t_2/t_{20} \quad \text{(dilatant flow)} \quad (16)$$

where t_1, t_{10} are the times for sample suspension and the suspending medium to pass through the first reservoir (a→b), and t_2, t_{20} are the times for passage through the second reservoir (b→c). If the liquid is Newtonian, the second and the third terms of Eq. (14) give the same value (within a relative error of 0.3%). If the two terms are not equal, the liquid is non-Newtonian and the flow is expressed by Eq. (15) or (16). In the case of Newtonian flow, the specific viscosity η_{sp} is

$$\eta_{sp} = \eta_{rel} - 1 \quad (17)$$

When the difference between t_1/t_{20} and t_2/t_{20} is within 1%, the values are averaged for the calculation of η_{sp} by Eq. (17). Such a slight shear dependence can be neglected and the sample can be regarded as a Newtonian fluid.

When the difference between the values of t_1/t_{10} and t_2/t_{20} is more than 5%, the flow time measurements are repeated by tilting the capillary and reducing the head difference until the difference is reduced to within 5% (Fig. 7). Shear dependence of the viscosity is thus reduced. Without any effects of orientation, plasticity, or dilatancy, both values become the same in a dilute suspension. When the difference between the values of $[\eta]$ for a→b and b→c is within 5%, the value for b→c, where the shear effect is smaller, is adopted or the three are averaged. (Simha calculated the axial ratio by assuming that Brownian motion makes the direction of particle

---

[11] D. Bensimon, *Phys. Rev. Lett.* **74**, 4754 (1995).

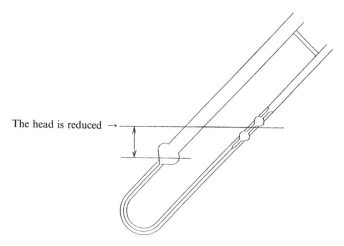

The head is reduced →

FIG. 7. Tilting of the viscometer.

axis isotropic. This condition is attained when the shear effect diminishes at a low flow velocity.)

Determination of Nondimensional Intrinsic Viscosity

Nondimensional intrinsic viscosity $[\eta]$ is determined by Eq. (8). As there is little inner aqueous phase in lipoplex, ϕ can be regarded to be ϕ_{net}. Reduced viscosity, $\eta_{red} = \eta_{sp}/\phi_{net}$, is then plotted against ϕ_{net}. ϕ_{net} is calculated by Eq. (7) from the amount of DNA and lipids in the sample. Although ρ of the lipoplex is unknown, $\rho = 1$ may be used. According to the Huggins equation, the plot is on a straight line in the low η_{red} region. When the obtained plots are not on a straight line, a least-squares line to the plots is drawn and extrapolated to $\phi_{net} = 0$. The intercept on the ordinate gives the nondimensional intrinsic viscosity (Fig. 8).

Results

Flow time data and viscosity are obtained with a conventional Ostwald capillary viscometer with an aqueous D-glucose solution (Table III). Almost the same viscosity data are obtained with the autominicapillary viscometer (Table IV) The flow times and viscosity data are shown in Table V and in Table VI with lipoplex suspension without fractionations. The intrinsic viscosity is determined to be 840 kg/g and the nondimensional intrinsic viscosity is 330. Corresponding axial ratios to 330 are obtained as 70 for a prolate ellipsoid and as 350 for an oblate ellipsoid. The coefficient of variation of the flow time (CV) seldom exceeds 0.1% and the data are

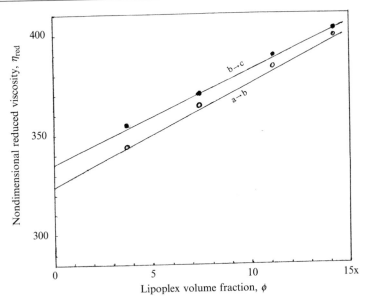

FIG. 8. Determination of nondimensional intrinsic viscosity, $[\eta]$, of DNA–lipid complex. *Sample:* Plant DNA, sonicated to 0.5 to 2 kbp, is complexed with distearyldimethylammonium chloride (DDAC, FW 575.5) at a molar ratio of 1:4, dissolved in 10 mM Tris-HCl buffer (pH 8.0). The complex has an FW of 2490 and a net density of 1.001. The final DNA concentration in the stock solution is 0.57 mN of the base unit. The stock soution is diluted to 0.149, 0.297, and 0.446 mN. *Measurements:* Viscosity is measured by automatic minicapillary viscometer at $40.00 \pm 0.01°$. *Data:* t_{10}, t_{20}: flow time (seconds) of 10 mM Tris-HCl buffer; t_1, t_2: flow time of lipoplex suspension; $\eta_{rel} = t_1/t_{10}$, t_2/t_{20} through the same reservoir; $\eta_{sp} = \eta_{rel} - 1$; $\eta_{red} = \eta_{sp}/\phi$; $\phi =$ volume of suspending lipoplex/volume of suspension.

stable. When the CV exceeds 0.3%, the data are deleted. This is often caused by the formation of a water-repellant layer on the inner surface of the capillary. The viscometer is rinsed with water and heated at $500°$ for 20 min. The data become stable again (Fig. 4). The relative viscosity from the second reservoir, t_2/t_{20}, b→c, is always larger than that from the first reservoir, t_1/t_{10}, a→b. This is considered to be caused by the orientation of particles to the direction of flow. As the difference is within 5%, the three values of $[\eta]$, a→b, a→c, and b→c, are averaged to give 330.

Discussion

According to the Sakurada equation [Eq. (13)], the shape parameter, $\alpha = 0$, means that the particles are rigid spheres, whereas $\alpha = 2$ means that the particles are straight filaments. Although Eq. (13) gives information

TABLE III

VISCOSITY OF AQUEOUS D-GLUCOSE SOLUTION AT 30°, BY OSTWALD VISCOMETER[a]

Sample	Weight fraction	Reservoir	Flow time, t (seconds)					AV	SD	CV	Relative viscosity
			1	2	3	4	5				
Water		a→b	118.38	118.51	118.48			118.46	0.07	0.0006	
		b→c	184.84	184.89	185.04			184.92	0.1	0.0006	
		a→c	303.22	303.4	303.52	303.45		303.4	0.13	0.0004	
1	0.01	a→b	123.57	123.3	123.35	123.26		123.37	0.14	0.0011	1.0414779
	0.01	b→c	192.91	193.12	192.93	193.01		192.99	0.1	0.0005	1.0436352
	0.01	a→c	316.48	316.42	316.28	316.27		316.36	0.1	0.0003	1.04273272
2	0.02	a→b	128.49	128.39	128.26	128.75		128.47	0.21	0.0016	1.08455272
	0.02	b→c	200.66	201.09	201.47	200.9		201.03	0.34	0.0017	1.08709916
	0.02	a→c	329.15	329.48	329.73	329.65	329.71	329.54	0.24	0.0007	1.08617902
3	0.03	a→b	133.93	134.44	134.62	134.3	134.43	134.34	0.26	0.0019	1.13411937
	0.03	b→c	209.49	210.11		210.49	210.32	210.1	0.44	0.0021	1.13616003
	0.03	a→c	343.42	344.55		344.79	344.75	344.38	0.65	0.0019	1.13507033
4	0.04	a→b	146.17	145.83	145.8	145.97	144.9	145.73	0.49	0.0034	1.23027267
	0.04	b→c	228.8	228.47	228.64		228.63	228.64	0.13	0.0006	1.23637724
	0.04	a→c	374.97	374.3	374.44		373.53	374.31	0.59	0.0016	1.23372803
5	0.05	a→b	146.3	146.46	146.16			146.31	0.15	0.001	1.23510707
	0.05	b→c	229.11	229.05	228.91			229.02	0.1	0.0004	1.23847721
	0.05	a→c	375.41	375.51	375.07			375.33	0.23	0.0006	1.23708996

Abbreviations: AV, Average; SD, standard deviation; CV, coefficient of variation.

[a] *Data:* t_{10}, t_{20}, t_{30}; flow time (seconds) of water; t_1, t_2, t_3; flow time of aqueous D-glucose solution from a to b (a→b), a→c, and b→c; relative viscosity $\eta_{rel} = t_1/t_{10}$, t_2/t_{20}, t_3/t_{30} through the same reservoir, e.g., a→b; weight fraction = weight of solute/weight of solution. *Temperature:* 30.00 ± 0.01°.

TABLE IV

VISCOSITY OF AQUEOUS D-GLUCOSE SOLUTION AT 30°, BY AUTOMINICAPILLARY VISCOMETER[a]

Sample	Weight fraction	Reservoir	Flow time (seconds)				AV	SD	CV	η_{rel}
			1	2	3	4				
Water		a→b	187.43		187.42	187.14	187.33	0.1646	0.0009	
		b→c	316.09		315.07	314.49	315.22	0.81	0.0026	
		a→c	503.52	502.33	502.49	501.63	502.49	0.7802	0.0016	
1	0.01	a→b	194.79	194.89	194.74		194.81	0.0764	0.0004	1.03991
	0.01	b→c	327.45	327.35	327.37		327.39	0.0529	0.0002	1.03862
	0.01	a→c	522.24	522.24	522.11	522.27	522.22	0.0714	0.0001	1.03925
2	0.02	a→b	204.28	203.58	203.72	204.17	203.94	0.3398	0.0017	1.08865
	0.02	b→c	339.53	339.66	340.38	339.74	339.83	0.3784	0.0011	1.07808
	0.02	a→c	543.81	543.24	544.1	543.91	543.77	0.3701	0.0007	1.08214
3	0.03	a→b	213.23	214.17	213.67	213.2	213.57	0.4555	0.0021	1.14006
	0.03	b→c	354.84	354.18	354.45	356.05	354.88	0.8257	0.0023	1.12583
	0.03	a→c	568.07	568.35	568.12	569.25	568.45	0.5487	0.001	1.13126
4	0.04	a→b	222.36	221.18	221.99		221.84	0.6035	0.0027	1.18424
	0.04	b→c	370.78	371.41	370.02		370.74	0.696	0.0019	1.17613
	0.04	a→c	593.14	592.59	592.01		592.58	0.5651	0.001	1.17928
5	0.05	a→b	235.83	236.04	235.51	236	235.85	0.2412	0.001	1.25898
	0.05	b→c	392.86	392.03	392.15	391.24	392.07	0.6636	0.0017	1.24381
	0.05	a→c	628.69	628.07	627.66	627.24	627.92	0.6179	0.001	1.2496

Abbreviations: AV, Average; SD, standard deviation; CV, coefficient of variation.

[a] *Data:* t_{10}, t_{20}, t_{30}: flow time (seconds) of water; t_1, t_2, t_3: flow time of aqueous D-glucose solution from a to b (a→b), a→c, and b→c; relative viscosity, $\eta_{rel} = t_1/t_{10}$, t_2/t_{20} through the same reservoir, e.g., a→b; weight fraction = weight of solute/weight of solution. *Temperature:* $30.00 \pm 0.01°$.

TABLE V

VISCOSITY OF SUSPENSION OF DNA–DDAC COMPLEX BY AUTOMINICAPILLARY VISCOMETER[a,b]

Sample	$c \times 10^3$	Reservoir	Flow time, t			AV	SD	CV	η_{rel}	η_{sp}	μ_{red}
			1	2	3						
Buffer (0.01 mM Tris–HCl)	0	a→b	158.88	159.08	159.14	159.03	0.136	0.00086			
		a→c	423.55	423.63	423.65	423.61	0.053	0.00012			
		b→c	264.67	264.55	264.51	264.58	0.083	0.00031			
1	0.147	a→b	179.32	179.41	179.14	179.29	0.137	0.00077	1.1274	0.1274	867
		a→c	478.24	479.11	478.57	478.64	0.439	0.00092	1.1299	0.1299	884
		b→c	298.92	299.7	299.43	299.35	0.396	0.00132	1.1314	0.1314	894
2	0.297	a→b	202.49	202.29	202.36	202.38	0.101	0.0005	1.2726	0.2726	918
		a→c	539.73	539.36	539.73	539.61	0.214	0.0004	1.2738	0.2738	922
		b→c	337.24	337.07	337.37	337.23	0.15	0.00045	1.2746	0.2746	925
3	0.446	a→b	226.37	226.89	227.39	226.88	0.51	0.00225	1.4266	0.4266	957
		a→c	604.76	605.88	606.94	605.86	1.09	0.0018	1.4302	0.4302	965
		b→c	378.39	378.99	379.55	378.98	0.58	0.00153	1.4324	0.4324	969
4 (stock)	0.57	a→b	248.88	249.12	249.02	249.01	0.121	0.00048	1.5658	0.5658	993
		a→c	665.05	664.7	664.22	664.66	0.417	0.00063	1.569	0.569	998
		b→c	416.17	415.58	415.2	415.65	0.489	0.00118	1.571	0.571	1002

Abbreviations: AV, Average; SD, standard deviation; CV, coefficient of variation.

[a] *Sample:* Plant DNA, sonicated to 0.5 to 2 kbp, is complexed with distearyldimethylammonium chloride (DDAC, FW 575.5) at a molar ratio of 1:4, and dissolved in 10 mM Tris–HCl buffer (pH 8.0). The complex has an FW of 2490 and a net density of 1.001. The final DNA concentration in the stock solution is 0.57 mN of the base unit. The volume fraction of lipoplex in the stock, Φ_v is 0.00142. The stock solution is diluted to 0.149, 0.297, and 0.446 mN. *Data:* t_{10}, t_{20}, and t_{30}; flow time (seconds) of 10 mM Tris–HCl buffer; t_1, t_2, and t_3; flow time of lipoplex suspension, from a to b (a→b), a→c, and b→c; $\eta_{rel} = t_1/t_{10}$, t_2/t_{20}, and t_3/t_{30} through the same reservoir, e.g., a→b $\eta_{sp} = \eta_{rel} - 1$. $\mu_{red} = \eta_{sp}/c$. *Temperature:* $40.00 \pm 0.01°$.

[b] Intrinsic viscosity is found to be $[\mu]_{a \to b} = 830$, $[\mu]_{a \to c} = 840$, $[\mu]_{b \to c} = 850$ from Fig. 2. Because these values are within a range of $\pm 3\%$, they are averaged to give 840.

TABLE VI

VISCOSITY OF SUSPENSION OF DNA–DDAC COMPLEX BY AUTOMINICAPILLARY VISCOMETER[a,b]

Sample	Φ		Flow time, t			AV	SD	CV	η_{rel}	η_{sp}	η_{red}
			1	2	3						
Buffer (0.01 mM Tris-HCl)	0	a→b	158.88	159.08	159.14	159.03	0.136	0.00086			
		a→c	423.55	423.63	423.65	423.61	0.053	0.00012			
		b→c	264.67	264.55	264.51	264.58	0.083	0.00031			
1	0.00037	a→b	179.32	179.41	179.14	179.29	0.137	0.00077	1.1274	0.127374	344
		a→c	478.24	479.11	478.57	478.64	0.439	0.00092	1.1299	0.129907	351
		b→c	298.92	299.7	299.43	299.35	0.396	0.00132	1.1314	0.13143	355
2	0.00074	a→b	202.49	202.29	202.36	202.38	0.101	0.0005	1.2726	0.272563	368
		a→c	539.73	539.36	539.73	539.61	0.214	0.0004	1.2738	0.273829	370
		b→c	337.24	337.07	337.37	337.23	0.15	0.00045	1.2746	0.27459	371
3	0.00111	a→b	226.37	226.89	227.39	226.88	0.51	0.00225	1.4266	0.42664	384
		a→c	604.76	605.88	606.94	605.86	1.09	0.0018	1.4302	0.430231	387
		b→c	378.39	378.99	379.55	378.98	0.58	0.00153	1.4324	0.432389	389
4 (stock)	0.00142	a→b	248.88	249.12	249.02	249.01	0.121	0.00048	1.5658	0.565751	398
		a→c	665.05	664.7	664.22	664.66	0.417	0.00063	1.569	0.56903	401
		b→c	416.17	415.58	415.2	415.65	0.489	0.00118	1.571	0.571	402

Abbreviations: AV, Average; SD, standard deviation; CV, coefficient of variation.

[a] *Sample*: Plant DNA, sonicated to 0.5 to 2 kbp, is complexed with distearyldimethylammonium chloride (DDAC, FW 575.5) at a molar ratio of 1:4, and dissolved in 10 mM Tris-HCl buffer (pH 8.0). The complex has an FW of 2490 and a net density of 1.001. The final DNA concentration in the stock solution is 0.57 mN of the base unit. The volume fraction of lipoplex in the stock, Φ, is 0.00142. The stock solution is diluted to 0.149, 0.297, and 0.446 mN. *Data*: t_{10}, t_{20}, and t_{30}: flow time (seconds) of 10 mM Tris-HCl buffer; t_1, t_2, and t_3: flow time of lipoplex suspension, from a to b (a→b), a→c, and b→c; $\eta_{rel} = t_1/t_{10}$, t_2/t_{20}, and t_3/t_{30} through the same reservoir, e.g., a→b. $\eta_{sp} = \eta_{rel} - 1$. $\eta_{red} = \eta_{sp}/\phi$ (nondimensional) *Temperature*: $40.00 \pm 0.01°$.

[b] Nondimensional intrinsic viscosity is found to be $[\eta]$ a→b = 325, $[\eta]$ a→c = 330, $[\eta]$ b→c = 335 from Fig. 1. Because these values are within a range of $\pm 3\%$, they are averaged to give 330. The prolate axial ratio is 70 and the oblate axial ratio is 440, from Table II.

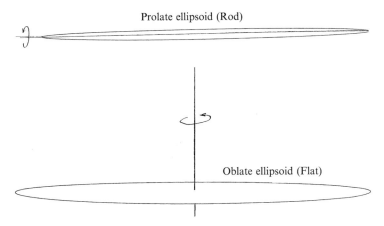

FIG. 9. Prolate ellipsoid with axial ratio 70 is more probable than oblate ellipsoid with axial ratio 350.

about the shape of suspending particles, determination of α requires many measurements of the dependence of $[\eta]$ on M. For this purpose, fractionation of the sample suspension and determination of ϕ of each fraction are needed.

The shape of liposomes and DNA–lipid complexes are also characterized by the axial ratio of an ellipsoid, using Eqs. (11) and (12). As Eqs. (11) and (12) hold irrespective of particle size, time-consuming fractionations and determinations of ϕ for each fraction are not necessary. Viscosity data of a suspension of DNA–DDAC lipoplex are shown in Tables V and VI. From the obtained nondimensional intrinsic viscosity of 330, axial ratios of 70 as prolate ellipsoids and 350 as oblate ellipsoids are determined. The latter means thin disks. Such a lipoplex has not been reported. A prolate ellipsoid with an axial ratio of 70 means that the lipoplex is fibrillar in shape, which is more probable (Fig. 9).

The formation of a lipoplex is thought to proceed in the following three steps.[12]

DNA + excess cationic lipid vesicles → complex 1 (rapid electrostatic process, within 1 s)

Aggregation of complex 1→ complex 2 (slow process by dispersion force within 1 h)

[12] V. Oberle, U. Bakowsky, I. S. Zuhorn, and D. Hoekstra, *Biophys. J.* **79,** 1447 (2000).

[13] D. D. Lasic, *J. Am. Chem. Soc.* **119,** 832 (1997).

Fusion of vesicles and collapse of the lipid film to form a multilamellar sheath around DNA (maturation)
→ lipoplex in equilibrium → complex 3 (slow process, hours or days at room temperature)

The time it takes for the solution to pass through the flow impinger is less than 1 s. Although this is enough for formation of complex 1, it is too short to complete the formation of complex 2. If the aggregates are dispersed in the solution with excess cationic lipid vesicles in some flow condition for a few hours, all the particles would become the final complex 3. The shape of the lipoplex is thought to depend on flow conditions during maturation. However, the relation between the flow condition and rheological shape parameter of the final lipoplex is not clarified at this time. A quantitative study to follow the time course of lipoplex maturation may be possible viscometrically, using Eqs. (11) and (12).

Concluding Remarks

An automated minicapillary viscometer, and its application to the characterization of liposomes and DNA–lipid complexes, is described. The viscosity of dilute small unilamellar vesicles obeys Einstein's viscosity equation for a suspension of nonattracting spherical particles. An average volume fraction can be determined by extrapolating the reduced viscosity-versus-concentration plot to zero concentration. For multilamellar vesicle suspensions and DNA–lipid complexes, the Einstein viscosity equation is extended to a suspension of ellipsoids by introducing the nondimensional intrinsic viscosity, which is related to the axial ratio, a/b. The nondimensional intrinsic viscosity $[\eta]$ is related to the axial ratio of an ellipsoid, which is applicable to low axial ratios (in the range $1 < a/b < 100$) by Eqs. (11) and (12).

A minicapillary viscometer, which uses less than 0.5 ml of sample suspension at a volume fraction of less than 0.1%, is automated by a laser sensor using an optical fiber. The procedure for the viscosity measurements and calculations are shown by an example using plant DNA–distearyldimethylammonium chloride complex at a charge ratio of 1:4, in which the amount of DNA is less than 250 μg. The prolate ellipsoidal axial ratio, a/b, is found to be 70. Determination of the shape parameter with a/b is found to be better than that with α of the Sakurada equation, because fractionation of particle size is not necessary. The shape of liposomes and bioactive polymer complexes is characterized quantitatively by the proposed method. The axial ratio of the DNA–lipid complex can be determined with less than 0.5 mg of DNA sample and corresponding lipid materials by using an automated minicapillary viscometer. The method described here may be utilized for quality control during large-scale production.

Acknowledgment

The author thanks Ms. Y. Sun (Dalian Institute of Chemical Physics, Chinese Academy of Science) for preparation of the lipoplex and for performing flow time measurements of viscosity, and Mr. Y. Takaoka (Tokyo Denki University) for automation of the minicapillary viscometer. The author is grateful to Prof. N. Düzgüneş (University of the Pacific), the late Dr. D. D. Lasic, and Prof. F. Szoka (University of California, San Francisco) for helpful discussions and suggestions.

[13]   Atomic Force Microscopy Imaging of Liposomes

By Jana Jass, Torbjörn Tjärnhage, and Gertrud Puu

Introduction

Atomic force microscopy (AFM) is one of a number of scanning probe microscopy techniques that image a sample surface with a sharp tip or probe. Binnings, Rohrer, Gerber, and Weibel designed the scanning tunneling microscope (STM). STM functions by creating a tunneling current between the sample and the probe, thus the sample must be conductive. Elegant engineering helped overcome the requirement for conductive samples, resulting in the development of the AFM, which has now become an important tool in imaging biological samples *in situ*. Images are obtained by scanning a sample surface with a sharp tip approximately 5–40 nm in diameter. The tip is maintained at a constant distance above the sample surface by repulsion of the electrons between the tip and sample surface. By placing the tip onto a flexible cantilever, the repulsion between the tip and surface is detected by the deflection of the cantilever as it approaches the surface. A laser reflecting off the cantilever end into a split-photodiode detector controls the piezo scanners via a computer system. An image is then produced by raster scanning across the surface, creating two separate images for the two scan directions. Figure 1 illustrates the design of the AFM. This technology allows imaging in any environment, liquid, air, or vacuum, thus providing high-resolution images of biological samples under biological conditions.

Atomic force microscopy has been used to visualize supported phospholipid membranes and, under some conditions, liposomes have also been

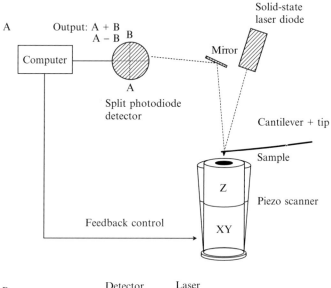

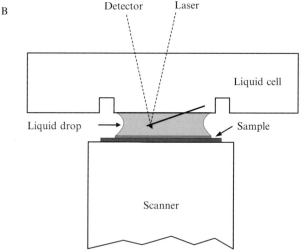

FIG. 1. A schematic representation of the atomic force microscope (A) and the liquid cell (B). (A) A laser light is focused onto the cantilever and into a split photodiode detector. The output from the detector goes to a computer that adjusts the piezo scanners accordingly and maintains the tip above the sample. This allows the tip to monitor the surface topography as it raster scans the sample. (B) In liquid, the laser light goes through a transparent glass liquid cell and reflects off the cantilever tip into the detector. The liquid makes a meniscus between the sample and the liquid cell. An O-ring usually fits between the grooves of the liquid cell and the sample to create a chamber; however, we do not use it, as it interferes with scanning large areas.

visible. Imaging can be made with the preparation in air, which is a simple technique.[1,2] However, the greatest advantage of AFM, as compared with electron microscopy, is that it permits visualization under physiological conditions, in water or buffer, with no prior manipulation of the liposomes. Another benefit is that processes can be monitored continuously at high resolution. There are three requirements for imaging liposomes in buffer. First, the liposomes must be attached to the surface, well enough so that the tip does not move them. Second, the experimental conditions, for example, the lipid composition of the liposomes, must be chosen to prevent the tendency of many types of liposomes to spontaneously break up and form a supported lipid bilayer. Third, as the tip may introduce artifacts when scanning the sample, one must try to avoid this by tuning the microscope properly and running control experiments. By selecting carefully the solid surface and surface modifications, and by selecting the composition of the liposomes and the conditions for imaging, valuable images of liposomes may be obtained. We address each of these items separately because there may be limited possibility to adjust one or a number of these parameters, thus optimizing the technique for individual samples.

Substrates

Selection of an appropriate substrate is important to obtain sufficient immobilization of the liposomes to the surface while avoiding rapid conversion to bilayer membranes in addition to having a flat background surface. The most common support used is mica.[1,3–6] Mica is a layered mineral that is cleaved easily to present a clean, atomically flat surface with a net negative charge. Muscovite mica consists of negatively charged SiO_4 molecules cross-linked by aluminum atoms with OH groups. A number of groups have also studied liposomes on modified mica, for example, to change the charge of the surface.[3,6–8] Other frequently used surfaces are silicon-based supports.[2,3,9] These substrates require stringent cleaning procedures to obtain a continuous oxide layer.

[1] S. Singh and D. J. Keller, *Biophys. J.* **60**, 1401 (1991).
[2] G. Puu, E. Artursson, I. Gustafson, M. Lundström, and J. Jass, *Biosens. Bioelectron.* **15**, 31 (2000).
[3] N. H. Thomson, I. Collin, M. C. Davies, K. Palin, D. Parkins, C. J. Roberts, S. J. B. Tendler, and P. M. Williams, *Langmuir* **16**, 4813 (2000).
[4] I. Reviakine and A. Brisson, *Langmuir* **16**, 1806 (2000).
[5] S. Kumar and J. H. Hoh, *Langmuir* **16**, 9936 (2000).
[6] Z. V. Leonenko, A. Carnini, and D. T. Cramb, *Biochim. Biophys. Acta* **1509**, 131 (2000).
[7] G. Luo, T. Liu, X. S. Zhao, Y. Huang, C. Huang, and W. Cao, *Langmuir* **17**, 4074 (2001).
[8] B. Pignataro, C. Steinem, H.-J. Galla, H. Fuchs, and A. Janshoff, *Biophys. J.* **78**, 487 (2000).
[9] J. Jass, T. Tjärnhage, and G. Puu, *Biophys. J.* **79**, 3153 (2000).

Silica, like mica, can be modified easily by chemical treatment, for example, silanization. However, we do not recommend modifying substrates to increase their hydrophobicity because that promotes the rapid conversion of liposomes to bilayers.[9]

Procedure for Cleaning Silicon Substrate

Polished (100)-oriented, boron-doped silicon wafers (Aurel, Landsberg, Germany) are cut into a size that fits onto the magnetic sample holders for the AFM used. The slides are cleaned in $NH_4OH-H_2O_2$-water (1:1:5, v/v/v) at 85° for 5 min, rinsed with water, and cleaned in $HCl-H_2O_2$-water (1:1:6, v/v/v) at 85° for an additional 5 min. The slides are rinsed with running ultrapure water for at least 30 min, after which they are either used immediately or stored for short periods of time in ethanol. Immediately before use, the slides are dried under a stream of nitrogen gas; once dried, the slides cannot be stored.

Liposomes

The method for liposome preparation—extrusion through filters, sonication, or detergent depletion—seems not to be crucial for success in AFM imaging. The problem is, not only liposomes composed of phosphatidylcholine, but many types of liposomes, have a tendency to form planar lipid structures on the substrate after attachment.[1,6] The aim must be to reduce the rate of conversion to a supported membrane. Lipid composition and liposome size are major factors that affect this tendency. Although not all researchers are able to alter these parameters to make the AFM study relevant to their specific work, it is important to realize that this may be the primary limitation in their use of AFM.

Liposome size has been implicated in stability at a surface. Smaller liposomes appear to be more stable, whereas larger vesicles spread into planar bilayers more rapidly. The rate of fusion for liposomes is believed to be determined by competition between the adhesive energy and the vesicle-bending energy.[10] Obviously this would also be influenced by lipid composition and ionic content of the buffer. Reviakine and Brisson have investigated this complex interdependence in some detail.[4] As an example of stable (nonfusing) conditions, it should be mentioned that sonicated unilamellar vesicles of an approximate diameter of 24 nm adsorb to most surfaces in an EDTA-containing buffer independent of lipid composition. Egg phosphatidylcholine unilamellar vesicles produced by extrusion

---

[10] P. Nollert, H. Kiefer, and F. Jähnig, *Biophys. J.* **69**, 1447 (1995).

through 30- or 50-nm pore filters also adsorb intact to surfaces in EDTA buffer.

Liposomes can also be polymerized for increased stability. Stanish *et al.* reported increased stability of polymerized phosphatidyl choline (PC) liposomes that did not fuse and retained their size and shape.[11] By incorporating disulfide functional groups into the lipids, the intact vesicles were able to bond to gold substrates and avoid aggregation. They were then able to image these vesicles using contact mode at relatively high forces of $0.75 \, N/m^2$.

AFM studies have also been done on liposomes containing not only phospholipids (sometimes also cholesterol) but also proteins,[2] the peptide gramicidin,[6] or glycolipids.[9] Such additives should make the liposomes slightly more hydrophilic and charged, which may help establish good attachment to the surface. The modifications are obviously not prohibitive for successful AFM imaging. This conclusion opens up exciting possibilities to monitor interactions, such as those between ligands and membrane receptors, in real time.

Finally, one way to avoid or reduce coalescence of attached liposomes to a planar membrane could be to keep the concentration of liposomes low. Results suggest that in some cases fusion may be stimulated if the attached liposomes are close enough to permit edge-to-edge contact on bilayer structures.[9] It may be that for various liposome compositions and sizes, a threshold concentration is required for optimum fusion to occur, which should then be avoided when visualizing intact attached liposomes.

Preparing Samples for Imaging in Air

To obtain images of liposomes in air, the vesicles must be able to withstand both washing with water and drying. Silicon slide substrates scored at 1 cm are cleaned and placed into a 4-ml spectrophotometer cuvette containing liposomes in 20 mM Tris-HCl (pH 7.4)–100 mM NaCl buffer–10 mM CaCl$_2$. An example of liposomes that have been used is as follows: mixed lipids (16 μM), consisting of 1,2-dipalmitoyl-*sn*-glycero-3-phosphocholine (DPPC), 1,2-dipalmitoyl-*sn*-glycero-3-phosphoethanolamine (DPPE), 1,2-dipalmitoyl-*sn*-glycero-3-phosphoglycerol (DPPG), and cholesterol (35, 25, 20 and 20 mol%, respectively), and biotin (0.023 mg/ ml)-labeled nicotinic acetylcholine receptor (Fig. 2). The liposomes are incubated overnight at room temperature and kept in suspension by stirring or recirculating with a peristaltic pump. The slides are washed

[11] I. Stanish, J. P. Santos, and A. Singh, *J. Am. Chem. Soc.* **123**, 1008 (2001).

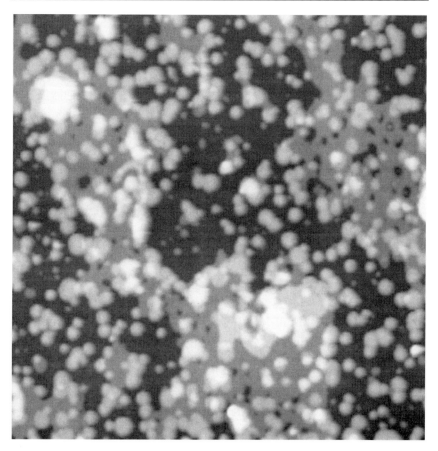

FIG. 2. Tapping Mode (TM)-AFM height image of acetylcholine receptor–containing liposomes in air. The membrane patches are visible underneath the liposomes. The height of the membrane structures is approximately 5.4–6.3 nm, whereas the liposomes range from 11 to 30 nm, suggesting that there are some collapsed liposomes, two double bilayers thick, and possibly some intact liposomes present on the surface. Proteoliposomes here appear much smaller than glycolipid-containing liposomes. The image parameters are as follows: scan size, 2 μm; scan rate, 0.65 Hz; cantilever oscillation frequency, 337 kHz; scan resolution, 512.

with Ca^{2+}-free buffer to remove any nonattached liposomes, followed by water to prevent salt crystal formation. The surface is dried under a stream of nitrogen, broken along the scored line to produce 1-cm^2 slides, and affixed to a magnetic sample holder. The samples are now ready for imaging.

Buffers for Liposomes and Imaging

Buffer components may contribute to liposome stability and aid in immobilizing them to the surface. The composition of the medium may also influence the level of noise in the imaging, with higher ionic- strength fluids resulting in greater noise.[6] As mentioned, a comparative study on liposomes was published and it included an investigation on the effect of various buffers.[4] The study showed that the presence of EDTA was beneficial for preserving liposome structure.

Calcium ions are commonly added to liposome buffer to aid in the formation of membrane structures; this tends to increase the rate of planar bilayer formation.[12] In addition, in the presence of approximately 5 mM calcium we have observed an increased level of noise when imaging the substrate. These buffers should therefore be avoided for AFM imaging.[4,9,13] If calcium is required for the preparation of liposomes, then addition of EDTA or EGTA may help stabilize the liposomes during imaging. During our AFM investigations of the effects of calcium on fusion we were able to image liposomes in Tris-HCl buffer (pH 7.4)–100 mM NaCl, with and without CaCl$_2$. Figure 3 illustrates the noisy background of a calcium-containing image (Fig. 3B) compared with Ca^{2+}-free Tris buffer (Fig. 3A). In addition, fewer liposomes were present and more bilayer structures were visible. Measurements of the bilayer heights were quite different between the two images, although they were from the same sample preparation (not the same region). Müller and Engel have reported previously that electrolyte concentration, pH, and salt affect the measured height of a sample.[13] These parameters then affect the interactions of both the support and sample with the tip and thus alter (often increase) the imaging forces.

Suggestions for Buffers or Solutions Used in AFM Imaging

20 mM Tris-HCl, pH 7.4, with 20 or 100 mM NaCl[2,4,9]
10 mM HEPES (pH 7.4), 2–4 mM EDTA, and 20–150 mM NaCl[4]
20 mM NaCl[5]
Water[6]

Preparing Samples for Imaging in Liquid

To conduct imaging in liquid, using a Nanoscope AFM, a liquid cell is a prerequisite and is commercially available from Digital Instruments (DI; Santa Barbara, CA). The liquid cell (Fig. 1B) is a square glass chamber

---

[12] G. Puu and I. Gustafson, *Biochim. Biophys. Acta* **1327,** 149 (1997).
[13] D. J. Müller and A. Engel, *Biophys. J.* **73,** 1633 (1997).

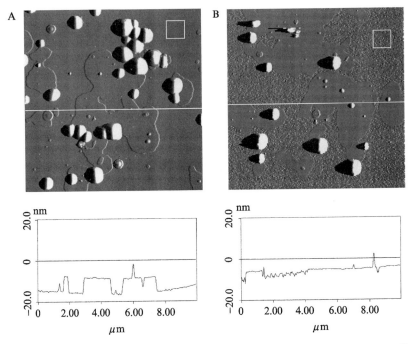

FIG. 3. The effect of calcium on AFM images of liposomes and bilayer structures on a silica surface. AFM images of glycolipid-containing liposomes in (A) Tris-HCl–NaCl buffer and after (B) the addition of calcium (4.5–7 mM). The most obvious effect seen is the increased noise of the silica surface, as illustrated by the cross-sections of each image and the mean roughness of the substrate calculated from within each box indicated on the image. The mean roughness values, Ra, are nearly twice as large for the image collected in the Ca^{2+}-containing sample buffer, with values of $R_a = 0.45$ and 0.75 nm for without and with Ca^{2+}, respectively. After the addition of Ca^{2+}, fewer liposomes, but more bilayer structures, are seen on the silica surface. The height of the membrane in (A) is 7.7 nm, whereas the height in (B) is much lower, 2.6 nm. The images were taken in TM-AFM and are presented in the error mode, and the x-section measurements were obtained from the height mode. The image parameters are as follows: scan size, 10 μm; scan rate, 1.5 Hz; cantilever oscillation frequency, 9.0 kHz; scan resolution, 256.

machined to hold the cantilever securely in place by a spring mechanism, with an inlet and outlet for liquid exchange. The liquid chamber may be sealed with an O-ring that sits in a groove within the cell and on the sample surface. The O-ring protects the scanners from liquid spills; however, it limits imaging and may affect the sample as it slides over the surface. Other microscope manufacturers may have different requirements for scanning in liquid environments that must be determined individually. The selected scanner should have sufficient range to image 10- to 20-μm areas to search for optimal regions and then reduce to 1- to 2-μm areas for imaging. The Digital Instruments J-scanner with the range of 125 μm is ideal for these experiments.

The liposome sample may be prepared either by overnight incubation as outlined for imaging in air or directly in the microscope. Slides prepared with attached liposomes from an overnight incubation are washed with Ca^{2+}-free buffer to remove nonattached liposomes and affixed carefully to the magnetic sample holder, without allowing the surface to dry before placement into the fluid cell. They are imaged in Ca^{2+}-free liposome incubation buffer. Slides may also be prepared directly in the microscope. A washed and cut silica surface affixed to an AFM magnetic sample holder is placed in the AFM. A drop of Ca^{2+}-free liposome incubation buffer is placed on the surface and imaging is initiated to obtain an image of a clean substrate. Scanning is halted while the liposome solution is injected into the buffer, using a Hamilton syringe inserted through the gap separating the liquid cell and the sample surface. This is possible because no O-ring is used in these experiments. The sample is incubated at room temperature in the microscope for several hours before imaging is resumed. The length of time for incubation is determined by the number of liposomes attached to the substrate. Because there is no mixing of the sample, few bound liposomes are found in the initial 2 h after addition of the liposomes.

Continuous imaging of samples is interesting only under buffer conditions in which fusion events are still occurring. The sample may be prepared as described; however, events may be clearer if the preparation is incubated directly in the microscope. This provides fewer liposomes at the surface and the initial events are still in progress. After initiating scanning and selecting a location on the surface, the scan parameters are optimized. The DI Nanoscope has a continuous scan command that collects consecutive images until a parameter is changed. It is important to remember that the consecutive images are produced from an alternating slow-scan direction that may be important in the interpretation of events. Other instruments may have similar commands or the user may need to initiate the Save command for each image.

Mode of AFM Imaging

Atomic force microscopy uses a sharp probe to scan the surface at a distance of 1 nm. There are two primary modes that exert different forces on the sample: contact mode and tapping or oscillation mode. Contact mode (CM-AFM) slides the tip across the sample surface while maintaining a constant distance and minimal force. The user manually and continually adjusts the instrument to maintain minimum forces on the sample while keeping the tip engaged. The forces involved here include lateral forces as the tip scans the surface. Contact mode is often used for imaging in liquid, where the force of the tip on the sample may be kept continuously at a minimum. The other most commonly used mode for imaging biological structures is TappingMode (TM-AFM; Digital Instruments) or oscillating mode, in which the tip taps the sample surface at its resonant frequency. Although this is believed to reduce lateral forces, point surface forces may be greater than in CM-AFM. This mode is often chosen to image samples in air because the effects of surface hydration are reduced. In our studies we have chosen this mode for imaging in liquid as well, using the cantilever with a nominal force constant of 0.35 N/m, which provides optimum tracking of the sample surface.

The choice of cantilever is based, first, on the mode of imaging and, second, on the minimal force required to maintain tracking of the surface. This is influenced by the sample properties in the imaging environment. From our experience, we obtained best results from standard or oxide-sharpened Si_3N_4 tips with a nominal force constant of 0.35 N/m for TM-AFM and 0.06 N/m for CM-AFM in liquid. For TM-AFM in air, standard "diving-board" silicon cantilevers are used. Scanning speed is optimized for the imaging mode and a pixel resolution of 256×256 is sufficient for the structures being imaged. Higher resolution does not provide more information and is not necessary.

It is often the case that some liposomes are associated with a surface but are not attached. Even if the forces are minimized the tip may move liposomes around the surface until they become attached. Figure 4 illustrates a two-frame sequence in which a liposome is moved nearly half the distance down the image, in the same direction as the slow scan. The movement is seen as a tip surface interaction and in the second frame the liposome is attached. Frequently, the tip will just move the unattached liposomes out of view, without creating a trail as seen in Fig. 4, suggesting that the tip has moved the liposomes out of view. Optimizing the parameters will not eliminate this effect, especially in early stages of liposome attachment.

Image Presentation and Interpretation

The scanned images can be viewed in different imaging presentations for optimizing visualization (Fig. 5). Changes in the deflection of the tip in CM and in the amplitude in TM may be used to create an image of the surface and this presentation is referred to as the "error image." The error image is often clearer for presentation and viewing of surface structures, especially if there is a large distance in the z direction. The height image is the distance the tip travels in the z direction and is necessary for any height

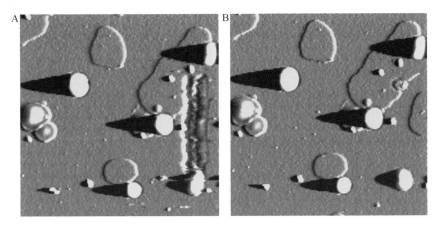

FIG. 4. Two sequential frames illustrating the tip moving a liposome over a silica surface. The image shows attached liposomes, partially flattened liposomes, and planar bilayers on the silica surface. In (A) a liposome in the lower right-hand corner is being moved down the frame, following the slow-scan direction. A path of tip interference indicates the liposome movement until the liposome stops and is imaged. (B) The slow-scan direction of the next image is up and the liposome is clearly visible in the location where it has stopped. The images of these glycolipid-containing liposomes were collected in TM-AFM in 20 mM Tris buffer (pH 7.4)–100 mM NaCl and are presented in the error mode. The image parameters are as follows: scan size, 6 μm; scan rate, 2.1 Hz; cantilever oscillation frequency, 8.9 kHz.

or distance measurements of the sample. Figure 5 illustrates the two presentation images for TM-AFM and the cross-sectional measurements obtained from the height image. The CM error image appears similar to that of TM and is therefore not illustrated here.

The AFM raster scans the surface and produces two images, one in each fast-scan direction. Images in both scan directions should be collected as this helps identify any possible artifacts or distortions. A common artifact observed is a multiple tip, which is often due to contamination of the scanning tip. It can be identified by small structures giving identical patterns and in the same order. Figure 6 illustrates an artifact that can alter seriously the interpretation of the image if not noticed. Careful inspection of the image identifies an artifact attributed to a double tip. If the artifact were not noticed it could be mistakenly assumed that there are two bilayers on top of each other rather than the correct interpretation of a single bilayer.

Example of Imaging Liposomes

Here we illustrate imaging of attached liposomes or liposome-like structures under different conditions and in real time. Again we want to stress that for specific applications, it will be necessary to

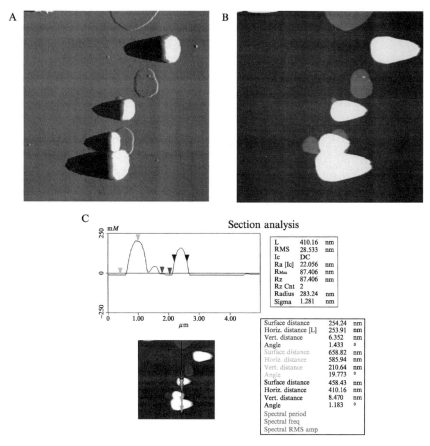

FIG. 5. TM-AFM images of liposomes and membrane disks presented as an (A) amplitude (error) and (B) height image. Cross-sectional analysis (C) is possible only from the height image; however, the structures are more clearly seen in the error mode, especially the liposomes, which are relatively large in the z direction compared with the membrane disks. The fast-scan directions are visible by the "shadow" produced during scanning and are from right to left for (A) and from left to right in (B). The cross-section of the image provides information about the image height, width, and surface distance between two surface structures. The glycolipid-containing liposomes were imaged by TM-AFM in 20 mM Tris buffer (pH 7.4)–100 mM NaCl. The image parameters are as follows: scan size, 5 μm; scan rate, 2.3 Hz; cantilever oscillation frequency, 9.0 kHz.

determine optimal conditions both for sample preparation and for tuning the AFM.

Unilamellar 200 to 400-nm liposomes are prepared by a detergent depletion technique.[14] They are composed of a lipid mixture containing DPPC, DPPE, DPPG, galactocerebroside type II, and cholesterol

A 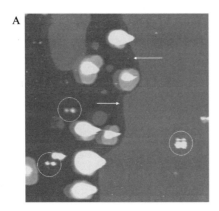   B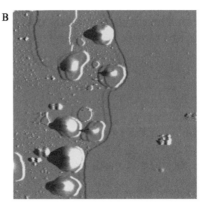

FIG. 6. (A and B) Tip artifacts in AFM images of liposomes and supported membranes. The image appears to show two bilayers on top of each other. Several features help identify that this is an artifact due to a double tip, most likely a result of contamination. First, the double membrane edge (arrows) always appears on the left side of the membrane and not randomly, as would be expected. The second indication is that smaller objects (circled) appear to come in pairs and always in the same orientation. This TM-AFM image is of proteoliposomes collected in 20 mM Tris buffer (pH 7.4)–100 mM NaCl and is presented in the error mode. The image parameters are as follows: scan size, 2.15 μm; scan rate, 2.54 Hz; cantilever oscillation frequency, 8.9 kHz.

(40:25:20:10:5, mol%). Some results use liposomes with no glycolipid and a higher DPPC content. DPPC, DPPE, and cholesterol are obtained from Avanti Polar Lipids (Alabaster, AL), whereas Sigma (St. Louis, MO) provides DPPG and the galactocerebroside. These mixed liposomes have been found previously to have a rather slow transfer to planar membranes and the galactocerebroside variant is especially suitable for continuous monitoring of liposome adhesion directly in the AFM cell. Samples are prepared as outlined previously. The liposome mixture is placed on a clean silica surface at a 1:5 dilution with buffer and the liquid cell is assembled.

The AFM is equipped with a DI J-scanner and a TM liquid cell. A small triangular cantilever is installed with a nominal force constant of 0.35 N/m. The tip is tuned to a rather low resonant frequency of approximately 7–9 kHz and an amplitude of approximately 500 mV. The scanning speed is optimized to 1.5–3.5 Hz and resolution is set to 256 × 256 pixels. Higher resolution requires more time to complete a scan without providing more information. Scanning is continued when there are sufficient numbers

[14] L. T. Mimms, G. Zampighi, Y. Nozaki, C. Tanford, and J. A. Reynolds, *Biochemistry* **20**, 833 (1981).

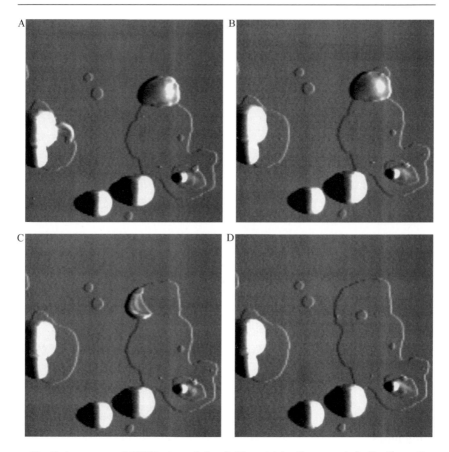

FIG. 7. A sequence of AFM images of glycolipid-containing liposomes in buffer illustrating real-time events of attached liposomes spreading into membrane structures on the silica surface. The images contain attached liposomes, membrane structures, and the spreading of one liposome to extend the membrane structure adjacent to it. The scan direction is either up or down, alternating with successive images: (A) down; (B) up; (C) down; (D) up. The total sequence was taken over 11 min and 20 s, with each image taking 170 s to collect. The images were collected in TM-AFM in 20 mM Tris buffer (pH 7.4)/100 mM NaCl and presented in the error mode. The image parameters are as follows: scan size, 5.0 μm; scan rate, 1.5 Hz; cantilever oscillation frequency, 9.0 kHz.

of liposomes attached to the surface, approximately 2 h after adding the liposomes.

To capture a sequence of events such as liposome spreading into membrane structures at the surface, it is necessary to select an area for imaging. The most promising regions should include both liposomes and planar bilayers, indicating that these events have been occurring. The capture of a sequence is random and may not be seen every time; however, persistence will result in images such as those illustrated in Fig. 7, in which a continual sequence of liposomes attaching, partially fusing, and finally forming a bilayer membrane has been collected over a period of 11 min and 20 s, with each frame taking 170 s to collect (Fig. 7).

Conclusion

Atomic force microscopy provides a unique possibility for visualizing liposomes *in situ* and any subsequent fusion event in real time. The primary considerations are that the liposomes be stable under the imaging conditions and be able to remain adsorbed sufficiently so that the tip does not move them around. Although this technique gives valuable information, it may not be possible in every case, and this must be determined for individual samples. Here we hope to have indicated the important factors required to image liposomes, and some examples in which this has been applied successfully.

[14] Interfacial Properties of Liposomes as Measured by Fluorescence and Optical Probes

By E. ANIBAL DISALVO, JESUS ARROYO, and DELIA L. BERNIK

Introduction: What Is the Lipid Interface?

The importance of liposomes is that they offer the possibility to be used as drug carriers and immunological adjuvants. Various attempts have been made to achieve appropriate formulations for oral vaccines, and pharmacological dispersions for intravenous administration.[1]

In this sense, the design of liposomes is a major activity within the field of biomimetic systems. The main objective is to obtain vesicles that may trap active substances and to ensure that they survive in the bloodstream or gastrointestinal tract to reach specific organs. There, they should be recognized by special receptors in order to be incorporated into the target cell.[2,3] This apparently simple sketch has faced many difficulties that

[1] G. Gregoriadis, *in* "The Theory and Practical Applications of Adjuvants" (D. E. S. Stewart-Tull, ed.), p. 145. John Wiley & Sons, New York, 1995.
[2] R. R. New, "Liposomes: A Practical Approach." IRL Press, New York, 1990.
[3] D. Lasic and F. Martin, "Stealth Liposomes." CRC Press, Boca Raton, FL, 1995.

have jeopardized the use of liposomes as delivery systems. Thus, much remains to be done to understand the physical chemistry of liposomes and how they change *in vivo*.

The design of liposomes as biomimetic systems must fulfill several requirements. Among them, the stability of the lipids in a bilayer is essential. This is important because the bilayer conformation is the structure that acts as a selective barrier for electrolytes and polar compounds. However, many compounds in the biological fluids may alter the bilayer permeability barrier by inducing a nonbilayer structure: the lamellar–hexagonal transition. This is achieved when liposomes are composed of lipids such as phosphatidylethanolamines and cardiolipins.[4]

Changes in membrane permeability may also be produced when lipids are still maintained in a bilayer conformation. In these cases, leakage can be induced by changes in the lateral packing of the bilayer. This can occur when the bilayer is subjected to stresses such as osmosis, partial dehydration, or the adsorption of amphiphilic compounds (detergents, bile salts, peptides, proteins, sugars, and amino acids).

In this regard, adsorption of compounds onto the liposome surface depends on the nature of the lipid interface. A lipid bilayer is an anisotropic structure composed of a hydrophobic core, corresponding to the acyl chain residues segregated from water, and the hydrophilic region of the polar head groups. The plane dividing these two regions is defined as the water–lipid interface, which seems to coincide with the plane containing the glycerol backbone. A region of around 20 Å thick from this plane toward the bulk water phase corresponds to the polar head groups and their hydration shells.[5,6] In addition, some penetration of water into the hydrocarbon region has been postulated as a result of the hydration of the carbonyls in the ester bonds between the glycerol backbone and the fatty acid residues.[7]

Properties of Lipid Interface

The organization at the membrane interface can, in principle, be ascribed to the conformation of the polar head groups, and the structure of the hydration shells around the chemical groups of the polar head

[4] P. Cullis and M. J. Hope, *in* "Biochemistry of Lipids and Membranes" (D. E. Vance and J. E. Vance, eds.), p. 25. Benjamin/Cummings, Menlo Park, CA, 1985.
[5] S. H. White and M. C. Wiener, *in* "Permeability and Stability of Lipid Bilayers" (E. A. Disalvo and S. A. Simon, eds.), p. 1. CRC Press, Boca Raton, FL, 1995.
[6] E. A. Disalvo and J. H. de Gier, *Chem. Phys. Lipids* **32,** 39 (1983).
[7] S. A. Simon and T. J. McIntosh, *Methods Enzymol.* **127,** 511 (1986).

components. This organization can be rationalized in terms of measurable properties such as packing, polarization, and charge distribution.

Packing is defined by the surface area occupied per lipid in the bilayer. Lateral packing is dictated by the head group–head group interaction and the cohesion of the hydrocarbon chains. The head-to-head interaction is affected by the hydration shell of the phospholipid chemical groups. This property can be measured directly in a monolayer by determining the lateral pressure versus area per lipid isotherm.[8] Curvature, hydration, and the intercalation of spacer molecules such as cholesterol may affect drastically the packing of the phospholipid head groups. This may result in the exposure of hydrophobic regions to the aqueous phase.

Polarization is directly related to the orientation of dipoles at the membrane interface and to the organization of water around the polar groups. This usually affects the local dielectric properties of the lipid interface. A property related to water polarization that can be measured in a monolayer spread on the air–water interface is the dipole potential.[8] This is built by the components of the dipoles of water and of the chemical groups of the membrane phospholipids normal to the membrane plane. It has been determined that carbonyls, the $P{=}O$ bond, and water are the main contributors to the dipole potential. Water bound to the charged and polar groups of the lipid molecules may be organized in a different way than in bulk water.[9,10] Around polar groups, polarization can arise from the orientation of the water dipoles toward net charges or by the formation of hydrogen bonds with polar residues. Around the methyl groups in phosphatidylcholines, water can form clusters displaying an icelike tetrahedral array.[11]

The replacement of polarized water molecules or its displacement by physical and chemical factors results in a drastic decrease in the dipole potential and in changes in the dielectric properties of the interface, and may affect the packing of the phospholipid groups. Several compounds that bind by hydrogen bonds to the carbonyl and phosphate groups, such as carbohydrates, polyphenols, detergents, and peptides, can affect the hydration of the phospholipids and hence the dipole potential. In addition, extrusion of water by osmosis can also affect packing and polarity by changes in curvature or in the hydration shell.

[8] H. Brockman, *Chem. Phys. Lipids* **73**, 57 (1993).

[9] W. Hübner and A. Blume, *Chem. Phys. Lipids* **96**, 99 (1998).

[10] S. B. Diaz, F. Lairion, J. Arroyo, A. C. Biondi de Lopez, and E. A. Disalvo, *Langmuir* **17**, 852 (2001).

[11] G. L. Jendrasiak and L. Hasty, *Biochim. Biophys. Acta* **337**, 79 (1974).

The charge distribution on the surface determines the surface potential. This potential is affected mainly by pH, ionic strength, and adsorbable ions such as Ca^{2+} and anions. Charges also produce a strong polarization of water molecules affecting the global polarity of the interface.[12]

Factors Affecting Surface Properties of Liposomes

The organization of the lipid groups and the water of hydration is inherent to the stabilization of the bilayer. Thus, in terms of surface science, it is important to define how packing, polarity, and electrical charges affect the surface properties of the lipid interface. In this regard we would like to center our discussion on the changes of liposome surfaces when they are affected by physical chemical variables such as osmotic stress, curvature, and phase coexistence. Various methodologies have been employed to obtain information about the lipid interface, both static and dynamically.

Electrical methods, such as electrical mobility, are useful to determine the net charge density on the surface by calculating the zeta (ζ) potential.[12] Dipole potential can be measured in bilayer lipid membranes and in monolayers spread on an air–water interface. Direct assessment of the $C{=}O$ and $P{=}O$ groups can be made by Fourier transform infrared spectroscopy (FTIR).[9,10]

In comparison, fluorescence methods can provide information about the packing of lipids in a bilayer and the polarization of the interface at a mesoscopic level.

The advantage of this methodology is that it may provide indications about the solvent properties of the region, regarding the hydrophobic or hydrophilic nature of the compounds interacting with the membrane, and about the polarization of the interface.

The membrane interface can be considered a surface solution composed of the polar head groups immersed in the hydration layer.[13] Thus, the solvent properties of the interface can be evaluated by considering the factors that may alter the water activity at the interface. For instance, the water activity at the interface can be changed if water can be dragged partially by an osmotic process and by the adsorption of solutes from the bulk water solution that may compete with bound water. In addition, water activity at the interface would be determined by the arrangement of the water dipoles in discrete regions around the area occupied by each lipid molecule. Thus, if a process such as ionization or dimerization depends

[12] S. McLaughlin, *Annu. Rev. Biophys. Chem.* **18**, 113 (1989).
[13] E. A. Evans and R. Skalak, "Mechanics and Thermodynamics of Biomembranes." CRC Press, Boca Raton, FL, 1980.

on the solvent properties, this can be used to evaluate the solvent properties of the interface if equilibrium is achieved between species confined to the interface. A measure of the change in surface properties may be obtained by determining the fluorescence properties of fluorophores that may be located in this region of the membrane.

Surface Fluorophores and Dyes

For these purposes, fluorescence methodologies are adequate because the quantum yield of the fluorophore depends on the solvent environment. As we are interested in the interfacial properties, fluorophores should be sensitive to changes in the polarity and packing in the polar region near the water–lipid interface. Some interfacial fluorophores are shown in Fig. 1, and their sensitivities to changes in polarity, surface potential, and packing are listed in Table I.

Merocyanine 540 partitions as a dimer at the outer plane of the interface and as a monomer in the region between the first methylenes of the acyl chains.[14,15] Laurdan (6-dodecanoyl-2-dimethylaminonaphthalene) locates at the inner part of the membrane at the level of the ester carbonyls at the glycerol backbone.[16,17] Dansyl groups (2-acetyl-1-[O-(11, 5-dimethylaminonaphthalene-1-sulfonyl)amino undecyl]-sn-glycero-3-phosphocholine) at the glycerol level, and Prodan (6-propionyl-2-dimethylaminonaphthalene) on the external plane of this region.[18]

Variations in the dipole potential of a membrane can be determined by recording the fluorescence emission of di-8-ANEPPS [1-(3-sulfonatopropyl)4-β[2-(di-n-octylamino)-6-naphthyl]vinyl)pyridinium betaine], as the ratio of two excitation wavelengths.[19] The behavior obtained with this probe can be compared with that found in direct measures of dipole potential in monolayers.

Merocyanine Partition between Lipid and Bulk Water Phase

Merocyanine 540 (MC540) (Fig. 1) is a lipophilic fluorescent dye that binds to the surface of plasma and liposome membranes. MC540 can be used quantitatively to study phospholipid packing and membrane phases

[14] P. I. Lelkes and I. R. Miller, J. Membr. Biol. 52, 1 (1980).
[15] L. Sikurova and R. Frankova, Studia Biophys. 140, 21 (1991).
[16] T. Parasassi and E. Gratton, Photochem. Photobiol. 57, 403 (1993).
[17] T. Parasassi and E. J. Gratton, J. Fluoresc. 5, 59 (1995).
[18] H. Rottemberg, Biochemistry 31, 9473 (1992).
[19] R. J. Clarke and D. J. Kane, Biochim. Biophys. Acta 1323, 223 (1997).

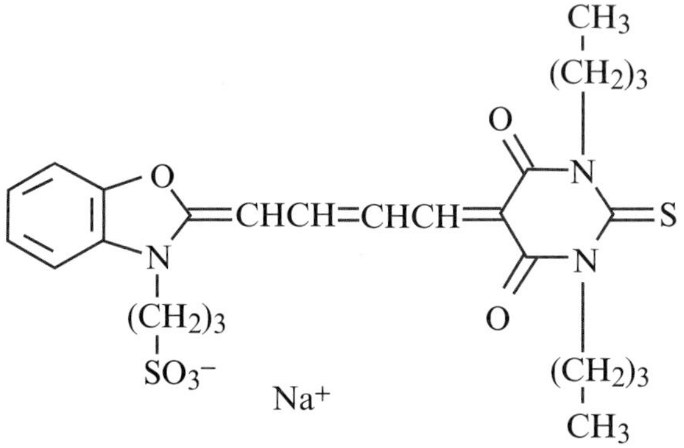

Merocyanine 540

Laurdan

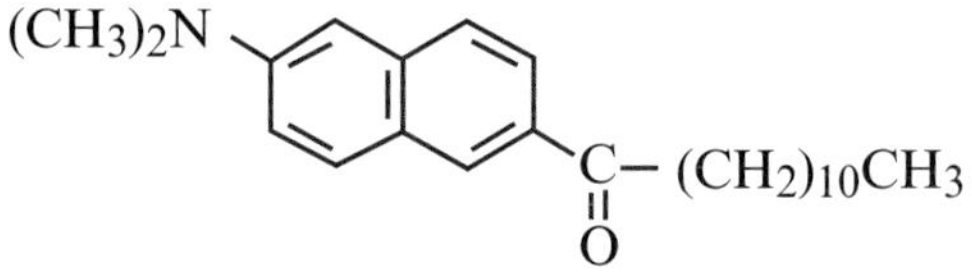

Dansyl phosphatidylcholine

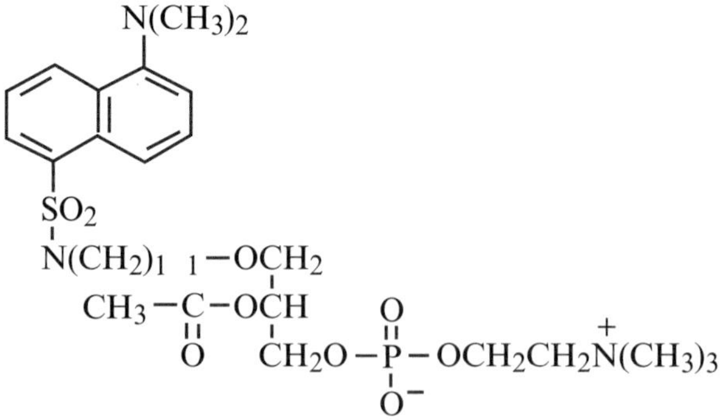

FIG. 1. Molecular formula of MC540, Laurdan, and Dansyl derivatives.

with lipid vesicles by measuring the absorbance spectra at the visible range or the fluorescence properties. Thus, it is a convenient compound, because relevant information can be obtained in a standard spectrophotometer. In addition, comparative fluorescence measurements with higher sensibility can be carried out.

TABLE I
CHANGES IN PACKING AND POLARITY AS MEASURED BY MC540, LAURDAN, PRODAN, AND DANSYL DERIVATIVES

Fluorescent probe	Anchored to membrane by	Probe location	Packing phosphates	Polarity phosphates	Packing carbonyls	Polarity carbonyls
Laurdan	Covalent nonpolar residue	Glycerol backbone toward the hydrocarbon phase	Slight increase	Not sensitive	No change	Not sensitive
Dansyl	Covalent nonpolar residue	Glycerol region	Increase	Decrease	Decrease	Slight decrease
Prodan	Covalent nonpolar residue	Glycerol region toward the water phase	Increase	Not sensitive	Decrease	Not sensitive
Merocyanine	Partition	External water phase	No effect or increase	Not sensitive	Decrease	Not sensitive

A solution of MC540 (2.5×10^{-3} M; Molecular Probes, Eugene, OR) in 10 mM Tris buffer, pH 7.3, is used as a stock solution. Aliquots are added to the vesicle dispersion in a cuvette (lipid concentration, 4 mM) to a final concentration of 10^{-7} to 10^{-6} M. Fluorescence quantum yield (Φ) is measured with a solution of MC540 in ethanol as reference ($\lambda_{ex} = 530$ nm, $\Phi_f = 0.15$ in ethanol at 25°, $\Phi_f = 0.6$ in dimyristoylphosphatidylcholine unilamellar vesicles at 5°).[20,21]

Merocyanine dissolved in water shows two peaks: one at 500 nm and another at 530 nm. The first peak has been ascribed to the dimer in water and the second to monomers in water. Further aggregates in water can be detected by the appearance of a peak at 450 nm. When dissolved in solvents of lower polarity, such as ethanol (dielectric constant, 24.3), MC540 shows a prevalent peak at 570 nm due to monomers. Thus, the partition of monomers between a nonpolar phase and a polar phase can be obtained by determining the ratio between the peaks at 570 and 530 nm.[20,21] Merocyanine 540 adsorbs to lipid membranes as a monomer (absorbance maximum at 570 nm) and as a dimer (absorbance maximum at 530 nm).[14] The monomer is the predominant species when the membrane is in the fluid state and the dimer is the predominant species when it is in the gel state. Characteristic spectra of MC540 in the presence of dipalmitoylphosphatidylcholine (DPPC) liposomes in the gel state and in the liquid crystalline state are shown in Fig. 2.

Surface Properties at Phase Transition as Determined by Merocyanine 540 Absorbance and Fluorescence

The properties of MC540 to adsorb in media of various dielectric properties have been used to determine changes in the phase state of the membrane, more particularly at the membrane interface. Merocyanine spectra can be taken after adding an aliquot of a stock solution to a liposome dispersion in order to obtain a 1:100 dye:lipid ratio. A control sample of vesicles of the same concentration without the dye is taken as a reference to eliminate turbidity contributions. At this probe:lipid ratio, the values of absorbances at 570 nm and at 530 nm can be taken from spectra of Fig. 2 run at different temperatures. The plot of the 570:530 absorbance ratio as a function of temperature shows an upward shift at the phase transition temperature of DPPC, reaching a stationary value in the fluid phase (Fig. 3). The fluorescence emission of MC540 monomer irradiated at 570 nm shows

[20] D. L. Bernik and E. A. Disalvo, *Chem. Phys. Lipids* **82**, 111 (1996).
[21] P. F. Aramendia, M. Krieg, C. Nitsch, E. Bittersman, and S. E. Braslavsky, *Photochem. Photobiol.* **48**, 187 (1988).

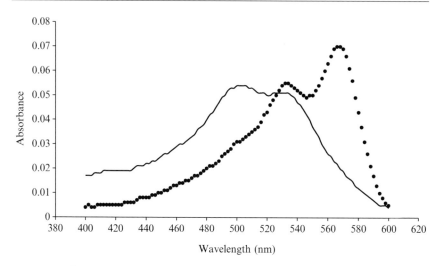

Fig. 2. Spectra of MC540 in gel (—) and fluid (-•-) membranes.

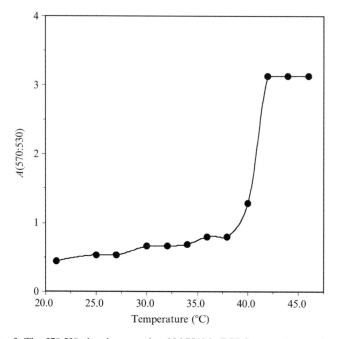

Fig. 3. The 570:530 absorbance ratio of MC540 in DPPC versus temperature.

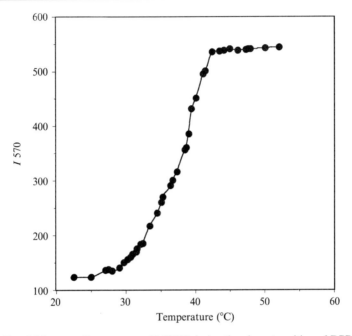

FIG. 4. Monomer fluorescence of MC540 during the phase transition of DPPC.

an abrupt change at 40–41°, which corresponds to the gel–liquid crystalline transition temperature. A complementary behavior can be obtained following the changes in absorbance at 530 nm (Fig. 4). Thus, changes in the 570:530 absorbance ratio and the fluorescence intensity report information about the changes taking place in the bilayer interface when a gel–liquid crystalline transition occurs.

Interfacial Properties Derived from Equilibrium Processes at Interface

The equilibrium distribution of monomers and dimers at the lipid–water interface is described schematically in Fig. 5, where the characteristic absorption and fluorescence wavelength of each species in the two media are denoted. The practical difficulty that affects the experimental determination of the interfacial properties of a bilayer is that amphiphilic probes such as MC540 partition between the adjacent aqueous solution and the membrane. The use of this type of probe has the disadvantage that species in the aqueous phase may interfere with the signal corresponding to the dye species in the membrane (see equilibrium described in Fig. 5). At the other

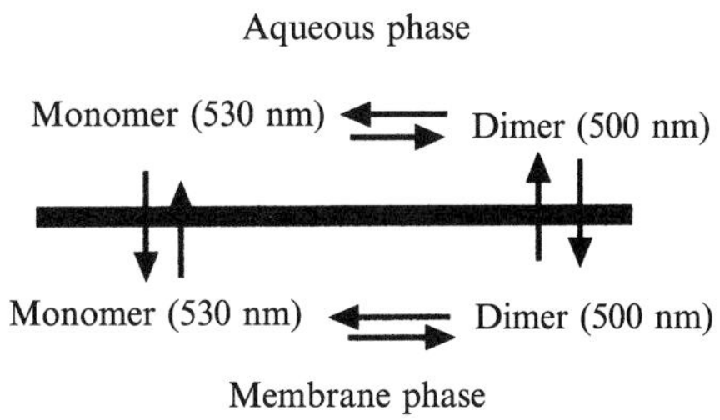

Aqueous phase

FIG. 5. Equilibrium of the various species of MC540 between the lipid and aqueous phases.

extreme, the location of fluorescent moieties anchored covalently to a hydrocarbon residue, such as Laurdan. Prodan, or Dansyl probes (see Fig. 1 and Table I), are strongly dependent on the nonpolar residues. In addition, they may be reporting properties of the solvent cage around them, which is only an indirect evaluation of the solvent properties of the water hydrating the membrane phospholipids. Thus, it is worthwhile to reconsider the use of probes that may partition freely between the interface and the bulk solution to obtain insights into the solvent properties of the interface.

The possibility that the 570- and 530-nm signals may report information about the properties of the interface requires the absence of dye species in the aqueous phase. This consists of preparing the liposome sample in a way in which monomer in the membrane can dimerize in the membrane interface, avoiding the other two steps described in the equilibrium of Fig. 5. A methodological procedure to achieve this point is described below.

After successive passages across the phase transition temperature (see Fig. 3), MC540 can be incorporated totally into the lipid membrane as a monomer, having an absorbance maximum at 568–570 nm. When the sample is cooled below the transition temperature, part of the monomer population dimerizes, giving a peak also at 530 nm. If the dye concentration is low enough, all the dye remains in the membrane phase, that is, the peak at 530 nm makes a low contribution of monomers in water. In this condition, any changes produced in the 570:530 ratio can be related to changes of the dimer–monomer equilibrium at the membrane level.

Absorption spectra can be obtained in a double-beam spectrophotometer controlling the temperature with a thermocouple within ±0.1°. Samples for a complete series of assays should be from the same batch of a stock vesicle dispersion. A baseline between 400 and 600 nm is determined with each sample of lipid dispersion without MC540, which is then subtracted from the spectra obtained with the same lipid sample in the presence of MC540 at the various dye:lipid ratios. The corresponding spectra are corrected by taking as zero the absorbances at 600 nm of the MC540–vesicle suspensions. This procedure should be followed to avoid errors due to the difference in turbidity and scattering effects, when the absorption values obtained with the MC540 concentrations assayed are low. At a high dye:lipid ratio, species in the aqueous solution can mask the presence of species in the membrane. For instance, a high absorbance at 530 nm due to monomers in water can hide the peak at 570 nm. In addition, in membranes in the gel state, the peak at 500 nm due to dimers in water can also be significant. The limit point of the MC540:lipid ratio above which this occurs can be detected by the absence of an isosbestic point.[20]

Determination of Dimerization Constant of MC540 at Membrane Interface

The dimerization constant of MC540 is defined by $K_d = [D]/[M]^2$. An increase in K_d denotes a decrease in the monomer concentration [M] in the membrane at the expense of an increase of the dimer concentration [D] per unit area of vesicles. As shown in Fig. 2, MC540 partitions into the fluid membranes as a monomer and in gel membranes as a dimer. Therefore, an increase in K_d may reflect several concurrent changes such as an increase in the packing of the bilayer at the level at which the monomer is partitioning or a decrease in the hydration of the interface. In addition, as is shown below, these changes may affect the membrane surface potential that can be determined by appropriate fluorophores.

The monomer and dimer concentrations of MC540 at the membrane can be calculated for each dye:lipid ratio from the total absorbance (A) of the sample per unit of optical path at a given wavelength:

$$A(\lambda) = \varepsilon_\lambda^M[M] + \varepsilon_\lambda^D[D] \tag{1}$$

where ε_λ^M and ε_λ^D represent the molar absorption coefficients of the monomer and the dimer species at 570 and 530 nm, respectively.[15]

Equation (1) includes monomer [M] and dimer [D] concentrations. The mass balance of MC540 in the volume dispersion is

$$C = [M] + 2[D] \tag{2}$$

where C is the total analytical concentration of MC540, expressed in moles per liter of solution of the sample of constant lipid concentration.

Reordering Eq. (2) and sustituting in Eq. (1), relation (3) can be obtained:

$$[M] = \{A_\lambda - (\varepsilon_\lambda^D/C/2)\}/\{\varepsilon_\lambda^M - \varepsilon_\lambda^D/2\} \tag{3}$$

and from the mass balance [Eq. (2)] dimer concentrations can be calculated as

$$[D] = (C - [M])/2 \tag{4}$$

Under conditions in which all the probe present in the solution is bound to the membrane either as a monomer or as a dimer, an equilibrium between monomer and dimer such as $2M \leftrightarrow D$ in the lipid phase with an apparent dimerization constant

$$K_{dapp} = [D]/[M]^2 \tag{5}$$

can be defined. Thus, a logarithmic plot of Eq. (5),

$$\log[D] = 2\log[M] + \log K_{dapp} \tag{6}$$

should give a straight line of slope 2. A typical curve is shown in Fig. 6. A slope of 2 denotes that the equilibrium is between the species in the interface, with negligible contribution of species in the bulk aqueous phase.

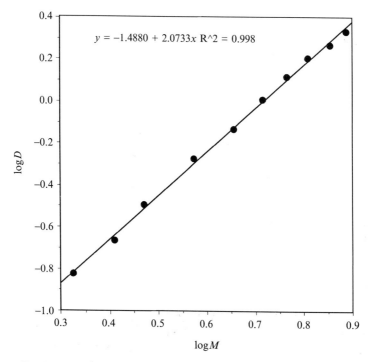

$y = -1.4880 + 2.0733x$ R^2 = 0.998

FIG. 6. Plot of log$M$ versus log$D$ under the conditions described in text.

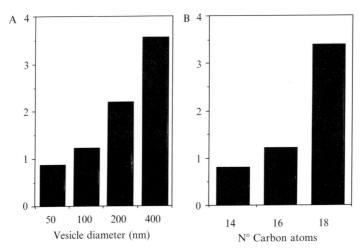

FIG. 7. Dimerization constant (K_d) of MC540 (A) as function of vesicle diameter in vesicles of dipalmitoylphosphatidylcholine; and (B) in vesicles of 100 nm as a function of lipid chain length.

The absorbance data for various merocyanine:lipid ratios (i.e., various MC540 concentrations) will fit Eq. (6) in a range of MC540:lipid ratios in which two MC540 species, and no more than two, are present in the system. Data of K_{dapp} are reported normalized by the total lipid concentration present in the dispersion.

Changes in Dimerization Constant due to Packing

It is well known that the increase in lateral pressure in a monolayer promotes an increase in packing. The limit is the collapse pressure at which the monolayer is destroyed and partly transformed into bilayers. This collapse pressure is lower with the increase in chain length. That is, longer chains are packed more tightly at comparable lateral pressures.

The dimerization of MC540 increases with chain length (Fig. 7) in 100-nm–diameter vesicles, denoting that variations in the dimerization constant are proportional to the increase in packing. In Fig. 7, the dimerization constants for vesicles of dipalmitoylphosphatidylcholine of increasing diameters are shown. Interestingly, it can be observed that similar surface properties (as measured with MC540) can be obtained with vesicles of 18:0 distearoylphosphatidylcholine (DSPC), 100 nm in diameter, or with vesicles of 16:0 dipalmitoylphosphatidylcholine (DPPC), 400 nm in diameter. Similar conclusions can be made with sonicated vesicles of DPPC and 100-nm-diameter vesicles of 14:0 dimyristoylphosphatidylcholine (DMPC).

Changes in Packing Induced by Osmotic Stress

The procedure to incorporate MC540 at the dye:lipid ratios described previously allows work with monomers in a gel membrane in the absence (or undetectable amount) of dye in the aqueous solution. The measure of the dimerization constant in osmotically stressed liposomes with respect to dimerization under isosmotic conditions can be used to monitor the relative degree of packing of the lipid interface when vesicles are subjected to osmotic shrinkage.[22] When 100-nm vesicles of DPPC are subjected to increasing osmotic pressure by adding polyethylene glycol (PEG, MW 10,000) to the external solution, the K_d increases as shown in Fig. 8A. The degree of packing is given by the relative change in the dimerization constant at each osmotic pressure with respect to the value corresponding to the isosmotic state for each vesicle size.

The degree of packing is shown to be linear with the logarithm of the osmotic pressure. The slopes of these straight lines are a function of the vesicle diameter. The effect of osmotic pressure on packing decreases with vesicle size (Fig. 8B). This denotes that, for a given osmotic pressure, large vesicles experience less change in area. The increase in packing produced in small sonicated vesicles is much greater than in 100-, 200-, and 400-nm vesicles. As shown in Fig. 8B, this behavior falls into the trend observed for vesicles of larger diameters. Small vesicles are considered more disordered than large vesicles as a consequence of their increased curvature. This is congruent with the fact that the outer area per lipid in a sonicated small vesicle is 72 Å^2, whereas in large vesicles it is 64 Å^2. Thus, they are more susceptible to increased packing under hypertonic stress.

Packing and Polarity Changes Measured with MC540 and Fluorophores Anchored to Lipid Phase

Fluid state bilayers subjected to hypertonic stress have a lower incorporation of monomers of MC540, as shown by the decrease in absorbance at 570 nm. The opposite is found in bilayers in the gel state.[23] Thus, merocyanine inserted in a position near the polar–apolar boundary could be used as a marker of structural changes taking place in that region when

[22] J. Arroyo, A. C. Biondi de Lopez, D. L. Bernik, and E. A. Disalvo, *J. Colloid Interface Sci.* **203**, 106 (1998).

[23] E. A. Disalvo, A. M. Campos, E. Abuin, and E. A. Lissi, *Chem. Phys. Lipids* **84**, 35 (1996).

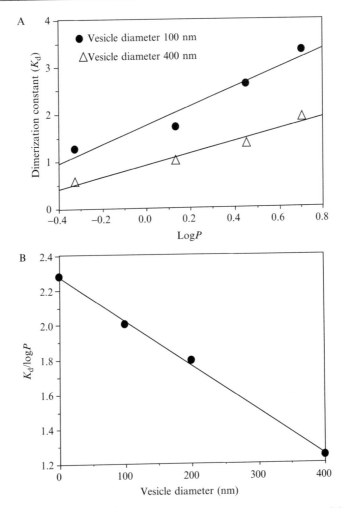

FIG. 8. (A) The log K_d versus osmotic pressure for vesicles of different size. (B) Slope of $K_d/\log P$ versus vesicle size.

osmotic shrinkage is induced either on a gel or a fluid membrane. The relative increase in the dimerization constant of MC540 induced by shrinkage can be ascribed to an extrusion of the dye monomer from the membrane phase caused by the pressure increase. As the work of membrane deformation depends on the activity of water at the interface,[13] this would be related not only to the amount of water molecules confined in the region but also to the degree of structure of such water molecules. The dimer:monomer ratio, reported in the absence of species in aqueous solution as a dimerization constant (K_d) of the dye species in the membrane, depends on the membrane phase properties. In this condition, the dimerization constant is within the order of that estimated for penthanol, for which the dielectric constant value is 14.[24] Thus, subtle changes in the degree of packing of the phospholipids of liposomes subjected to various levels of stress exploit the fact that the dye can be present as a monomer or as a dimer according to the dielectric properties of the medium in which it is dissolved. This property can be quantified by the K_d values. It is expected from this that MC540 can give information about changes in polarity and solvent relaxation when the hydration sites are perturbed by dehydrating agents.

Changes in the hydration of the carbonyls and phosphates can be perturbed by phloretin and trehalose.[25,26] In Table I we compare the results obtained with MC540 with those obtained with fluorophores that are anchored to the lipid phase by a hydrocarbon moiety and which are sensitive to polarity and packing changes (see Fig. 1 and Table I). The fluorescent probes Laurdan, Prodan, and Dansyl-hexadecylamine are added to the organic solvent used to prepare the lipid film before hydration to make liposomes. These samples can be analyzed as multilamellar dispersions or as unilamellar vesicles after extrusion through membranes of various pore sizes.

Steady-State Fluorescence Anisotropy Determinations Performed with Dansyldihexadecylamine

Steady-state fluorescence anisotropy can be calculated by means of the relation[27]

$$<r> = (I_{vv} - GI_{vh})/(I_{vv} + 2GI_{vh})$$

where I_{vv} and I_{vh} are the intensities of the fluorescence emission at the wavelength maximum obtained with the analyzer parallel or perpendicular, respectively, to the direction of polarization of the excitation beam. In all experiments, the light scattering of the samples without probe must be checked, and maintained at negligible levels when measuring the excitation:emission ratio in the absence of the fluorophore. I_{vv} and I_{vh} values must be corrected for phototube sensitivity by the geometric factor

[24] D. L. Bernik and R. M. Negri, *J. Colloid Interface Sci.* **203**, 97 (1998).
[25] J. H. Crowe, L. M. Crowe, and D. Chapman, *Science* **223**, 701 (1984).
[26] G. L. Jendrasiak, R. L. Smith, and T. J. McIntosh, *Biochim. Biophys. Acta* **1329**, 159 (1997).
[27] G. Bhaskar Dutt, M. Ameloot, D. L. Bernik, R. M. Negri, and F. C. De Schryver, *J. Phys. Chem.* **100**, 9751 (1996).

(G). $G = I_{hh}/I_{hv}$ is a correction factor accounting for the polarization bias in the detection system. In this case, I_{hv} and I_{hh} correspond to the intensities obtained with the excitation polarizer in the horizontal position (h) and the emission polarizer in the vertical position (v) or the horizontal position (h), respectively.

The polarity measurements performed by measuring the excitation and emission spectra of the probe in each system include the wavelength of the corresponding intensity maximum of the spectra in the equation

$$S = 1/\lambda_{max\ ex} - 1/\lambda_{max\ em}$$

where $\lambda_{max\ ex}$ and $\lambda_{max\ em}$ are expressed in centimeters. The higher the value of S, the higher the polarity of the probe microenvironment.

Dipolar Relaxation Determined by Laurdan and Prodan Probes

Prodan and Laurdan probes have the same chromophore group. The difference is that Laurdan has an alkyl chain covalently linked to the chromophore, anchoring the probe deep inside the bilayer–water interface (Fig. 1).

The generalized polarization parameter (GP) for both probes is calculated by means of the equation

$$GP = (I_B - I_R)/(I_B + I_R)$$

where I_B and I_R are the fluorescence intensities measured under conditions in which wavelength B and R are observed, respectively.[16] The emission GP spectra are constructed by calculating the GP value for each emission wavelength as follows:

$$GP_{cm} = (I_{390} - I_{340})/(I_{390} + I_{340})$$

where I_{390} and I_{340} are the intensities at each emission wavelength, from 420 to 510 nm, obtained using fixed excitation wavelengths of 390 and 340 nm, respectively. The excitation GP spectra are constructed in the same way but from the excitation spectra, using

$$GP_{ex} = (I_{440} - I_{490})/(I_{440} + I_{490})$$

where I_{440} and I_{490} are the intensities at each excitation wavelength, from 320 to 400 nm, obtained using fixed emission wavelength of 440 and 490 nm, respectively. The upper and lower limit values found for GP_{ex} and GP_{em} are 0.7/−0.3 and 0.3/−0.4, for the gel- and liquid crystalline-phase state, respectively, at least for phosphatidylcholines.

When the GP of Laurdan is dependent on the excitation wavelength, it may imply the coexistence of phase domains.[17] In general, the higher the

GP values, the lower the dipolar relaxation sensed by the probe, indicating a more rigid environment. Dansyl and Prodan probes, both located at the glycerol backbone, show that the phosphate region increases packing on the binding of phloretin. In contrast, trehalose, which binds to the phosphates but also affects the carbonyls, decreases packing when measured with the same probes (Table I).

Packing and Dipole Potential

The distribution of MC540 between the aqueous phase and the membrane phase has also been used extensively as an indicator of membrane potential.[28] A significant increase in the signal of the monomer in the membrane phase has been measured on membrane depolarization.[29] In addition, compounds such as trehalose and phloretin bind to the membrane, causing a decrease in the water of hydration.[30,31] The increase in monomer partition induced by trehalose can therefore be interpreted as a decrease in membrane potential due to the replacement of water molecules bound to the $C=O$ groups exposed to water. These results are in agreement with the decrease in dipole potential measured in monolayers.[31] However, the results with phloretin, which changes the dipole potential without changing the MC540 spectra, would suggest that this fluorophore is sensitive to packing but not to the dipole potential. Thus, it may be possible that the changes in MC540 would result from the decrease in packing promoted by the trehalose bound to the carbonyl parallel to the membrane plane.[31,32]

Under isosmotic conditions, the dimerization constant of MC540 increases with vesicle diameter (Fig. 7). Thus, lipids in larger vesicles are more densely packed, and therefore more organized. Figure 9 shows the changes in surface membrane potential as measured with di-γ-ANEPPS in the fluid and the gel state. It is observed that it increases in gel-state membranes with the increase in K_d.

As K_d is a function of packing due to chain length increase (Fig. 7), or water extrusion (Fig. 8), the increase in surface membrane potential can be ascribed to a decrease in area per dipole. This may be due to an increase in the relative proportion of polarized water in relation to nonpolarized water, as seems to be the case when the membrane goes from the fluid to

[28] T. Aiuchi and Y. Kobatake, *J. Membr. Biol.* **45,** 233 (1979).

[29] W. Stilwell, S. R. Wassall, A. C. Dumaual, W. D. Ehringer, C. W. Browning, and L. J. Jenski, *Biochim. Biophys. Acta* **1146,** 136 (1993).

[30] D. L. Bernik, D. Zubiri, E. Tymczyszyn, and E. A. Disalvo, *Langmuir* **17,** 6438 (2001).

[31] M. C. Luzardo, F. Amalfa, A. M. Nuñez, S. B. Diaz, A. C. Biondi de Lopez, and E. A. Disalvo, *Biophys. J.* **78,** 2452 (2000).

[32] S. Diaz, F. Amalfa, A. C. Biondi de Lopez, and E. A. Disalvo, *Langmuir* **15,** 5179 (1999).

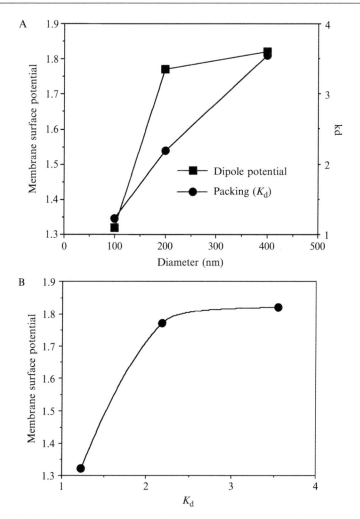

Fig. 9. (A) Dimerization constant and membrane surface potential as measured with di-8-ANEPPS as a function of vesicle size of phosphatidylcholines in the gel state. (B) Correlation between membrane surface potential and packing as measured by K_d.

the gel state. In this case, the dipole potential in a monolayer increases from 450 to 510 mV for DMPC.

The osmotic shock produced by osmolytes such as sucrose increases packing, as shown in Fig. 7, with an increase in interface polarization. This is congruent with the increase in dipole potential found in monolayers in the presence of sucrose.[31] In addition, it has been calculated that the hypertonic shock of liposomes increases the surface potential by approximately 7 mV at 25°.[33]

Final Remarks

Liposome surface properties can be measured with optical probes that may dimerize in the membrane interface. The dimerization constant may report information about changes in packing in the interface as a consequence of the chain–chain interaction, curvature (vesicle size), phase state, and osmotic state. Changes in packing are congruent with changes in polarity and membrane surface potential measured with fluorescent probes located at the interface. This could be ascribed to changes in the distribution of dipoles as measured by the dipole potential in monolayers.

[33] E. A. Disalvo, *in* "Surface Chemistry and Electrochemistry of Membranes" (T. S. Sørensen, ed.), p. 837. Marcel Dekker, New York, 1999.

[15] Thermotropic Behavior of Lipid Mixtures Studied at the Level of Single Vesicles: Giant Unilamellar Vesicles and Two-Photon Excitation Fluorescence Microscopy

By LUIS A. BAGATOLLI

Introduction

The effect of temperature on phase equilibria in lipid mixtures has been studied for almost 30 years and phase diagrams have been constructed, using both theoretical and experimental approaches.[1–12] As a consequence, important thermodynamic information is currently available for several lipid mixtures.

[1] J. H. Ipsen and O. G. Mouritsen, *Biochim. Biophys. Acta* **944,** 121 (1988).
[2] K. Jørgensen and O. G. Mouritsen, *Biophys. J.* **69,** 942 (1995).
[3] A. G. Lee, *Biochim. Biophys. Acta* **413,** 11 (1975).
[4] B. R. Lentz, Y. Barenholz, and T. E. Thompson, *Biochemistry* **15,** 4529 (1976).
[5] S. Mabrey and J. M. Sturtevant, *Proc. Natl. Acad. Sci. USA* **73,** 3862 (1976).
[6] P. W. M. Van Dijck, A. J. Kaper, H. A. J. Oonk, and J. De Gier, *Biochim. Biophys. Acta* **470,** 58 (1977).
[7] K. Arnold, A. Lösche, and K. Gawrisch, *Biochim. Biophys. Acta* **645,** 143 (1981).
[8] M. Caffrey and F. S. Hing, *Biophys. J.* **51,** 37 (1987).
[9] E. J. Shimshick and H. M. McConnell, *Biochemistry* **12,** 2351 (1973).

One interesting temperature region in the lipid mixture phase diagram is that corresponding to the formation of a stable lipid domain structure, which was discovered in 1977.[13] This phenomenon, which involves lipid phase separation and is characterized by the coexistence of ordered (tightly packed) and disordered (loosely packed) lipid phases in the bilayer, plays a central role in the stabilization of multicomponent vesicles and in the fission of small vesicles after budding.[14] Several studies on lipid domain structure and phase connectivity in bilayers that display phase coexistence were carried out using fluorescence recovery after photobleaching (FRAP).[15–19] The information extracted from all these different experimental techniques is useful to understand, starting from simple models, the possible occurrence of phase separation phenomena in complex systems such as cell membranes. It is important to note that the presence of phase coexistence in cell membranes is still a matter of controversy. However, accumulating experimental evidence favors the presence of laterally organized domains in biological membranes.[20–24]

Direct Visualization of Lipid Domain Coexistence in Bilayers

Even though lipid phase coexistence is well accepted in model systems (monolayers and bilayers), the direct visualization of lipid domains was successfully accomplished only in lipid films at the air–water interface, using fluorescence microscopy techniques.[24–28] Changes in lateral pressure

[10] B. Maggio, *Biochim. Biophys. Acta* **815**, 245 (1985).
[11] B. Maggio, G. D. Fidelio, F. A. Cumar, and R. K. Yu, *Chem. Phys. Lipids* **42**, 49 (1986).
[12] L. A. Bagatolli, B. Maggio, F. Aguilar, C. P. Sotomayor, and G. D. Fidelio, *Biochim. Biophys. Acta* **1325**, 80 (1997).
[13] C. H. Gebhardt, C. H. Gruler, and E. Sackmann, *Z. Naturforsch.* **32C**, 581 (1977).
[14] E. Sackmann and T. Feder, *Mol. Membr. Biol.* **12**, 21 (1995).
[15] W. L. C. Vaz, E. C. C. Melo, and T. E. Thompson, *Biophys. J.* **56**, 869 (1989).
[16] W. L. C. Vaz, E. C. C. Melo, and T. E. Thompson, *Biophys. J.* **58**, 273 (1990).
[17] T. Bultmann, W. L. C. Vaz, E. C. C. Melo, R. B. Sisk, and T. E. Thompson, *Biochemistry* **30**, 5573 (1991).
[18] P. F. F. Almeida, W. L. C. Vaz, and T. E. Thompson, *Biochemistry* **31**, 7198 (1992).
[19] V. Schram, H. N. Lin, and T. E. Thompson, *Biophys. J.* **71**, 1811 (1996).
[20] J. F. Tocanne, *Commun. Mol. Cell. Biophys.* **8**, 53 (1992).
[21] K. Simons and E. Ikonen, *Nature* **387**, 569 (1997).
[22] K. Jacobson and C. Dietrich, *Curr. Opin. Cell Biol.* **9**, 84 (1999).
[23] D. A. Brown and E. London, *Annu. Rev. Cell Dev. Biol.* **14**, 111 (1998).
[24] R. G. Oliveira and B. Maggio, *Neurochem. Res.* **25**, 77 (2000).
[25] R. M. Weis and H. M. McConnell, *Nature* **310**, 47 (1984).
[26] H. Möhwald, A. Dietrich, C. Böhm, G. Brezesindki, and M. Thoma, *Mol. Membr. Biol.* **12**, 29 (1995).
[27] K. Nag, J. Perez-Gil, M. L. F. Ruano, L. A. D. Worthman, J. Stewart, C. Casals, and K. M. W. Keough, *Biophys. J.* **74**, 2983 (1998).

induce the formation of coexisting lipid phases in lipid monolayers composed of pure lipids or lipid mixtures (either artificial or natural mixtures). The direct observation of this last phenomenon, using epifluorescence microscopy, generally involves the use of different fluorescent probes with preferential partitioning to one of the coexisting lipid phases (for details see section on Fluorescent Probes).

In the past few decades, direct visualization of lipid domains in bilayers at the phase coexistence region was achieved by electron microscopy techniques.[29] As was pointed out by Raudino, however, no direct and detailed knowledge of the shape and formation of domains in bilayers was available.[30] Actually, there are few experimental approaches that allow direct visualization of lipid domain formation (shape and dynamics) in lipid bilayers under the same experimental conditions as in classic approaches (such as, e.g., differential scanning calorimetry and fluorescence spectroscopy in a cuvette).

The aim of this chapter is to introduce an experimental approach that allows visualization of temperature-dependent lipid phase equilibria at the level of single vesicles, using two-photon fluorescence microscopy. Various methods to prepare giant unilamellar vesicles (GUVs) and novel results in the direct visualization of the lipid phase coexistence at the level of single vesicles, combining the particular properties of some fluorescent probes and two-photon excitation fluorescence microscopy, are discussed in this chapter.

Some Considerations about Giant Unilamellar Vesicles

To mimic the lipid lateral organization of the cell plasma membrane, GUVs are the proper choice as a model membrane system (mean diameter, ~20 μm). These "cell-sized vesicles" are attractive experimental systems to study the phase equilibria of pure lipid systems and lipid mixtures by fluorescence microscopy techniques, mainly because single vesicles can be observed under the microscope. Surprisingly, few studies based on fluorescence microscopy and GUVs have described lipid phase coexistence. In this sense, it is important to remark on the seminal contribution of Haverstick and Glaser, who achieved the first visualization of lipid domains in GUVs by fluorescence microscopy with digital image processing.[31] These

[28] R. Veldhuizen, K. Nag, S. Orgeig, and F. Possmayer, *Biochim. Biophys. Acta* **1408,** 90 (1998).

[29] E. Sackmann, *Ber. Bunsenges. Phys. Chem.* **82,** 891 (1978).

[30] A. Raudino, *Adv. Colloid Interface Sci.* **57,** 229 (1995).

[31] D. M. Haverstick and M. Glaser, *Proc. Natl. Acad. Sci. USA* **84,** 4475 (1987).

authors directly visualized Ca^{2+}-induced lipid domains in erythrocyte ghosts, GUVs formed of mixtures of phosphatidylcholine (PC) and acidic phospholipids, and GUVs formed from natural lipids from the erythrocyte membrane at constant temperature.[31] In addition, Glaser and co-workers also studied lipid domain formation caused by addition of proteins and peptides to GUVs.[32–34]

GUVs have been objects of intense scrutiny in diverse areas that focus on membrane behavior.[35] In particular, the advantages of direct observation of a single giant vesicle, using transmission and fluorescence microscopy, was well appreciated in areas such as chemistry, physics and biophysics. In this sense, an excellent review by Menger and Keiper[35] about GUVs and a book completely devoted to giant vesicles, edited by P. L. Luisi and P. Walde,[36] provide the reader with an overview of the initial stages in GUV research. For example, an intense study of membrane physics has been done with GUVs, in particular studies on the mechanical properties of model membranes.[35–42] These studies revealed the physical properties of the membranes through the calculation of elementary deformation parameters. Studies of lipid–lipid, lipid–protein, and lipid–DNA interactions were also performed with GUVs.[43–56] Still, the enormous potential capabilities of this model system are in an early stage of development.

[32] D. M. Haverstick and M. Glaser, *Biophys. J.* **55**, 677 (1989).

[33] M. Glaser, *Commun. Mol. Cell. Biophys.* **8**, 37 (1992).

[34] L. Yang and M. Glaser, *Biochemistry* **34**, 1500 (1995).

[35] F. M. Menger and J. S. Keiper, *Curr. Opin. Chem. Biol.* **2**, 726 (1998).

[36] P. L. Luisi and P. Walde, "Giant Vesicles." John Wiley & Sons, London, 2000.

[37] E. Evans and R. Kwok, *Biochemistry* **21**, 4874 (1982).

[38] D. Needham, T. M. McInstosh, and E. Evans, *Biochemistry* **27**, 4668 (1988).

[39] D. Needham and E. Evans, *Biochemistry* **27**, 8261 (1988).

[40] P. Meléard, C. Gerbeaud, T. Pott, L. Fernandez-Puente, I. Bivas, M. D. Mitov, J. Dufourcq, and P. Bothorel, *Biophys. J.* **72**, 2616 (1997).

[41] P. Meléard, C. Gerbeaud, P. Bardusco, N. Jeandine, M. D. Mitov, and L. Fernandez-Puente, *Biochimie* **80**, 401 (1998).

[42] E. Sackmann, *FEBS Lett.* **346**, 3 (1994).

[43] L. A. Bagatolli and E. Gratton, *Biophys. J.* **77**, 2090 (1999).

[44] L. A. Bagatolli and E. Gratton, *Biophys. J.* **78**, 290 (2000).

[45] L. A. Bagatolli and E. Gratton, *Biophys. J.* **79**, 434 (2000).

[46] L. A. Bagatolli, E. Gratton, T. K. Khan, and P. L. G. Chong, *Biophys. J.* **79**, 416 (2000).

[47] C. Dietrich, L. A. Bagatolli, Z. Volovyk, N. L. Thompson, M. Levi, K. Jacobson, and E. Gratton, *Biophys. J.* **80**, 1417 (2001).

[48] L. A. Bagatolli and E. Gratton, *J. Fluoresc.* **11**, 141 (2001).

[49] D. P. Pantazatos and R. C. MacDonald, *J. Membr. Biol.* **170**, 27 (1999).

[50] R. Wick, M. I. Angelova, P. Walde, and P. L. Luisi, *Chem. Biol.* **3**, 105 (1996).

[51] S. Sanchez, L. A. Bagatolli, E. Gratton, and T. Hazlett, *Biophys. J.* **82**, 2232 (2002).

[52] M. L. Longo, A. J. Waring, L. M. Gordon, and D. A. Hammer, *Langmuir* **14**, 2385 (1998).

General Considerations for Preparation of Giant Unilamellar Vesicles

In general, experimental conditions, such as ionic strength, pH, lipid composition, substrate on which to dry the lipid film, and addition of some sugars, seem to be critical parameters to obtain giant vesicles.[38,57–61] Nevertheless, there is no general agreement about "unique" conditions to obtain such vesicles, mainly because the mechanism of giant vesicle formation is still obscure. A major consequence of the lack of precise knowledge is that many different methods to obtain GUVs have been described.[38,57–70] For instance, low ionic strength (below 10 mM) in aqueous solution is required to successfully prepare giant vesicles.[38,58,59,61] Alternatively, as reported by Akashi *et al.*, physiological conditions can be used to obtain the giant lipid structures, using a percentage of charged lipids in the sample.[60] As a general rule two important conditions are required to prepare GUVs: (1) the temperature during vesicle preparation must be higher than the phase transition temperature of the lipids used to form the GUVs (this condition also operates in the formation of multilamellar, small unilamellar, and large unilamellar vesicles), and (2) to form large structures agitation of the samples during the vesicle formation must be prevented.

[53] P. Bucher, A. Fischer, P. L. Luisi, T. Oberholzer, and P. Walde, *Langmuir* **14,** 2712 (1998).

[54] M. I. Angelova, N. Hristova, and I. Tsoneva, *Eur. Biophys. J.* **28,** 142 (1999).

[55] J. Korlach, P. Schwille, W. W. Webb, and G. W. Feigenson, *Proc. Natl. Acad. Sci. USA* **96,** 8461 (1999).

[56] G. W. Feigenson and J. T. Buboltz, *Biophys. J.* **80,** 2775 (2001).

[57] A. Moscho, O. Orwar, D. T. Chiu, B. P. Modi, and R. N Zare, *Proc. Natl. Acad. Sci. USA* **93,** 11443 (1996).

[58] J. P. Reeves and R. M. Dowben, *J. Cell. Physiol.* **73,** 49 (1969).

[59] J. Käs and E. Sackmann, *Biophys. J.* **60,** 825 (1991).

[60] K. Akashi, H. Miyata, H. Itoh, and K. Kinosita, Jr., *Biophys. J.* **71,** 3242 (1996).

[61] M. I. Angelova and D. S. Dimitrov, *Prog. Colloid Polym. Sci.* **76,** 59 (1988).

[62] L. Yang and M. Glaser, *Biochemistry* **35,** 13966 (1996).

[63] M. I. Angelova and D. S. Dimitrov, *Faraday Discuss. Chem. Soc.* **81,** 303 (1986).

[64] D. S. Dimitrov and M. I. Angelova, *Prog. Colloid Polym. Sci.* **73,** 48 (1987).

[65] D. S. Dimitrov and M. I. Angelova, *Bioelectrochem. Bioenerget.* **19,** 323 (1988) [a section of *J. Electroanal. Chem.* constituting Vol. 253].

[66] M. I. Angelova, S. Soléau, P. Meléard, J. F. Faucon, and P. Bothorel, *Prog. Colloid Polym. Sci.* **89,** 127 (1992).

[67] M. I. Angelova and D. S. Dimitrov, *Prog. Colloid Polym. Sci.* **76,** 59 (1988).

[68] D. D. Lassic, *Biochem. J.* **256,** 1 (1988).

[69] M. Winterhalter and D. D. Lassic, *Chem. Phys. Lipids* **64,** 35 (1993).

[70] F. M. Menger and M. I. Angelova, *Acc. Chem. Res.* **31,** 789 (1998).

Comparison of Methods Used to Generate GUVs

One of the first experimental protocols reported (and the most popular) to generate giant vesicles was based on the exposure of dried lipid films to aqueous solution for a long time (up to 24 h) at temperatures above the lipid phase transition.[58] Some modifications of this last technique were reported to improve the yield of single-walled liposomes.[34,38,59,60] In all cases, giant vesicles with a mean diameter of approximately 15 to 20 μm can be obtained. Another technique based on organic solvent evaporation in aqueous solution was presented by Moscho *et al.* to prepare GUVs.[57] One of the advantages of this last method, compared with that presented above, is the short time required to obtain the vesicles (a few minutes). In this case vesicles up to 50 μm in diameter are obtained.[57] One of the drawbacks of these two techniques is that the yield of GUVs is low (no more than 20%). This last fact makes finding a single-walled vesicle under the microscope a time-consuming process. In addition, the vesicle size distribution obtained by these methods is broad and a significant number of multilamellar vesicles as well as other types of lipid structures (such as lipid tubes or vesicles with internal structures) are observed at the end of the formation process.[71]

An interesting technique to generate GUVs was introduced by Dimitrov and Angelova in 1986, consisting of hydration of the dried lipid films above to the lipid phase transition in the presence of electric fields.[63–67] In this last case, a high yield of GUVs (~95%), with a narrow size distribution (5 to 60 μm) with respect to the distributions obtained by the other methods previously described, is achieved in a reasonable period of time (60 to 90 min).[67,72] Figure 1 shows pictures of the GUV-generating chamber designed in our laboratory,[44,45] together with an image of the border of the Pt wire after vesicle formation. Once the vesicles are formed the AC field is turned off and the vesicles remain adsorbed to the Pt wires. This last fact is remarkable and allows a single vesicle to be monitored for long periods of time without vesicle drifting.[43–46] The many advantages of the electroformation method over other protocols have been discussed.[71] In particular, Bagatolli *et al.*[71] compare different preparation techniques and emphasize the necessity of evaluating the characteristics of the entire sample. In general, all the studies already made on such samples focus only on the GUVs, and do not consider the rest of the sample.[71] In Fig. 2, two-photon excitation fluorescence images of various lipid structures obtained by various GUV formation protocols are presented for comparison.

[71] L. A. Bagatolli, T. Parasassi, and E. Gratton, *Chem. Phys. Lipids* **105,** 135 (2000).
[72] L. Mathivet, S. Cribier, and P. F. Devaux, *Biophys. J.* **70,** 1112 (1996).

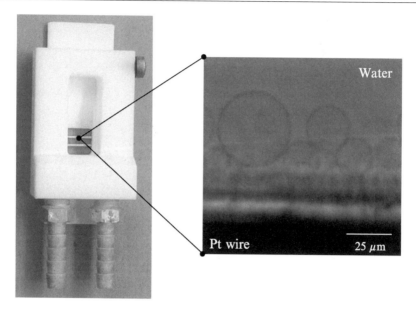

FIG. 1. Homemade chamber for GUV electroformation. The chamber is made of Teflon (inert material). The bottom of the unit has the same dimensions as a standard microscope slide. The separation between the Pt wires (diameter, 1 mm) is 3 mm (center to center) and the distance from the bottom of the chamber to the center of the Pt wires is minimal (100 μm in this case). The total cost of the equipment is relatively low, making this method efficient for generating GUVs. *Inset:* Transmission microscope image taken with a CCD color video camera (CCD-Iris; Sony) in an inverted microscope after vesicle formation.

Why Two-Photon Excitation Microscopy?

Two-photon excitation is a nonlinear process in which a fluorophore absorbs two photons simultaneously. Each photon provides half the energy required for excitation. The high photon densities required for two-photon absorption are achieved by focusing a high peak power laser light source on a diffraction-limited spot through a high numerical aperture objective.[73] Therefore, in the areas above and below the focal plane, two-photon absorption does not occur, because of insufficient photon flux. This phenomenon allows a sectioning effect without using emission pinholes as in one-photon confocal microscopy.[73]

Studies of lipid–lipid interactions in GUVs, using two-photon excitation[43–46,74] or conventional confocal[55,56] fluorescence microscopy, have been

---

[73] W. Denk, J. H. Strickler, and W. W. Webb, *Science* **248,** 73 (1990).

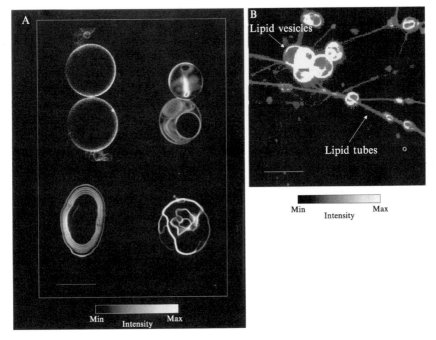

FIG. 2. (A) Fluorescence images of various lipid structures obtained by two-photon excitation microscopy. These lipid structures were obtained by different GUV formation protocols. *Top left:* Giant unilamellar vesicles obtained by the electroformation method. These vesicles are also visualized by the gentle hydration and solvent evaporation methods, but their occurrence is considerably reduced with respect to the electroformation method (see text). *Top right* and *bottom left:* Liposomes with internal structure obtained by the gentle hydration method. *Bottom right:* Liposomes with internal structure obtained by the solvent evaporation method. These structures occur frequently in samples obtained by the gentle hydration and solvent evaporation methods. (B) Two-photon excitation fluorescence image of lipid vesicles connected by lipid tubes. This image occurs frequently in samples obtained by the gentle hydration method. LAURDAN was used as the fluorescent probe. Scale bar: 20 μm.

reported. The direct visualization of lipid domain coexistence obtained by these last experimental approaches opened a fascinating new window on the topological features of the lipid bilayer when two different lipid phases coexist (see later). The advantages and disadvantages between using confocal or two-photon excitation microscopy have been well reviewed[73,75–77]; this chapter focuses instead on why two-photon excitation fluorescence

[74] K. Nag, J. S. Pao, R. R. Harbottle, F. Possmayer, N. O. Petersen, and L. A. Bagatolli, *Biophys. J.* **82,** 2041 (2002).

microscopy is used in the author's laboratory. An important advantage of two-photon excitation is the low extent of photobleaching and photodamage above and below the focal plane. This last point is crucial for experiments done with rapidly fading ultraviolet (UV) probes[78] such as 6-lauroyl-2-(N, N-dimethylamino)naphthalene (LAURDAN), 6-propionyl-2-(N, N-dimethylamino)naphthalene (PRODAN), or 1,6-diphenyl-1,3,5-hexatriene (DPH). For instance, the photobleaching effects on LAURDAN fluorescence (a favorite fluorescent molecule in the study of lipid–lipid interactions) are dramatic when using epifluorescence microscopy.[48] However, when using two-photon excitation, LAURDAN-labeled giant vesicles can be imaged for hours. The other important aspect in the author's experimental approach is the sectioning effect that allows observation of the vesicle surface without background contributions; this is essential to observe the formation of lipid domains (see later). This effect can also be accomplished by conventional confocal microscopy, using fluorescent probes that are excited in the visible wavelength range.[55,56] Another attractive advantage of two-photon excitation is that one wavelength can be used to excite many different fluorescent molecules simultaneously. Technically, this possibility can be exploited to perform multiple emission color experiments without changing the excitation wavelength.[76]

Fluorescent Probes

Several fluorescent probes have been reported to be useful in the study of the physical characteristics of lipid bilayers [see, e.g., the Molecular Probes (Eugene, OR) handbook], using mainly cuvette studies. In particular, to ascertain lipid domain coexistence in lipid bilayers or monolayers by fluorescence microscopy, fluorescent molecules that show different partition properties between the ordered and disordered phases are used.[24–27,44,45,55,56] In general, the fluorescence intensity images display the shape of the lipid domains, but additional techniques are necessary to

[75] P. T. C. So, T. French, W. M. Yu, K. M. Berland, C. Y. Dong, and E. Gratton, *in* "Fluorescence Imaging Spectroscopy and Microscopy" (X. F. Wang and B. Herman, eds.), Chemical Analysis Series, Vol. 137, p. 351. John Wiley & Sons, New York, 1996.

[76] B. R. Master, P. T. C. So, and E. Gratton, *in* "Fluorescent and Luminescent Probes," 2nd Ed., p. 414. Academic Press, New York, 1999.

[77] P. T. C. So, T. French, W. M. Yu, K. M. Berland, C. Y. Dong, and E. Gratton, *Bioimaging* **3**, 49 (1995).

[78] T. Parasassi, L. A. Bagatolli, E. Gratton, M. Levi, F. Ursini, W. Yu, and H. K. Zajicek, *in* "Confocal and Two-Photon Microscopy: Foundations, Applications and Advances" (A. Diaspro, ed.), p. 469. John Wiley & Sons, New York, 2001.

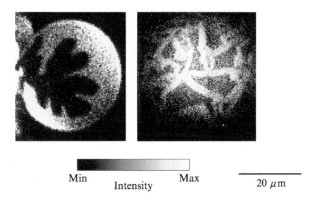

Min   Max
Intensity   20 μm

FIG. 3. Two-photon excitation fluorescence intensity images of N-Rh-DPPE-labeled GUVs composed of DPPE–DPPC (7:3, mol/mol) (*left*) and DLPC–DPPC (1:1, mol:mol) (*right*).

further evaluate the local physical characteristics of the observed lipid domains.[55,79]

In general, there is a tendency to classify fluorescent molecules that present different partition properties between coexisting phases [as in the case of Lissamine rhodamine B 1,2-dihexadecanoyl-*sn*-glycero-3-phosphoethanolamine (N-Rh-DPPE)] as gel- or fluid-like phase probes. Figure 3 shows a comparison between N-Rh-DPPE–labeled GUVs composed of dilauroylphosphatidylcholine and dipalmitoylphosphatidylcholine (DLPC–DPPC) and dipalmitoylphosphatidylethanolamine (DPPE)–DPPC at the phase coexistence temperature region. Even though it is well known from the phase diagram of the phospholipid binary mixture that both samples display gel/fluid-phase coexistence, it is difficult after examination of Fig. 3 to determine which domain corresponds to the gel or fluid phase in each sample. This last finding shows clearly that it is difficult to establish directly from the GUV fluorescent image the lipid domain phase state, using fluorescent probes that show only differential partition between the coexisting phases (such as N-Rh-DPPE),[44,48] and that it is not wise to generalize the affinity of a fluorescent molecule for the different lipid phases without a careful probe characterization. However, it is possible to circumvent this last problem by choosing particular fluorescent probes with particular fluorescence properties, as described in the next section.

[79] R. G. Oliveira and B. Maggio, *Biochim. Biophys. Acta* **1561,** 238 (2002).

Extracting Lipid Domain Phase State Information Directly from
Fluorescence Intensity Images

An easy way to extract phase state information directly from fluorescence intensity images is to take advantage of the particular fluorescent properties of LAURDAN. This molecule belongs to the family of polarity-sensitive fluorescent probes, first designed and synthesized by G. Weber for the study of dipolar relaxation of fluorophores in solvents, bound to proteins, and associated with lipids.[80-83] When inserted in lipid membranes, LAURDAN displays unique characteristics compared with other fluorescent probes,[48,84-88] namely (1) homogeneous probe distribution on lipid membranes displaying phase coexistence, (2) a phase-dependent emission spectral shift, that is, LAURDAN's emission is blue in the ordered lipid phase and greenish in the disordered lipid phase (this last effect is attributed to the reorientation of water molecules present at the lipid interface near the LAURDAN fluorescent moiety), and (3) parallel alignment of the electronic transition dipole of this molecule with the hydrophobic lipid chains in lipid vesicles.

Because LAURDAN is distributed homogeneously between the coexisting lipid phases in the membrane, the less ordered domains have a red-shifted emission compared with the more ordered domains.[43-45,48,84-88] This last fact can be observed easily by using a blue (or green) bandpass filter to obtain fluorescence intensity images at the equatorial region of the vesicle.[44-46,48] Figure 4 shows an image of LAURDAN-labeled GUVs composed of DPPE–DPPC, taken at the equatorial region of the vesicle with a blue bandpass filter. The high-intensity regions on the lipid bilayer correspond to the gel phase and LAURDAN is present in both coexisting domains. For comparison, a fluorescence image of N-Rh-DPPE–labeled GUVs composed of DPPE–DPPC is included in Fig. 4. Note that in this case the fluorescent molecule is excluded from the gel phase because this particular probe shows different partitioning between the coexisting phases.[44,45,48]

[80] G. Weber and F. J. Farris, *Biochemistry* **18,** 3075 (1979).
[81] R. B. MacGregor and G. Weber, *Nature* **319,** 70 (1986).
[82] T. Parasassi, F. Conti, and E. Gratton, *Cell. Mol. Biol.* **32,** 103 (1986).
[83] M. Lasagna, V. Vargas, D. M. Jameson, and J. E. Brunet, *Biochemistry* **35,** 973 (1996).
[84] T. Parasassi and E. Gratton, *J. Fluoresc.* **5,** 59 (1995).
[85] T. Parasassi, E. Krasnowska, L. A. Bagatolli, and E. Gratton, *J. Fluoresc.* **8,** 365 (1998).
[86] L. A. Bagatolli, E. Gratton, and G. D. Fidelio, *Biophys. J.* **75,** 331 (1998).
[87] T. Parasassi, G. De Stasio, G. Ravagnan, R. M. Rusch, and E. Gratton, *Biophys. J.* **60,** 179 (1991).
[88] T. Parasassi, G. De Stasio, A. d'Ubaldo, and E. Gratton, *Biophys. J.* **57,** 1179 (1990).

LAURDAN   N-RH-DPPE

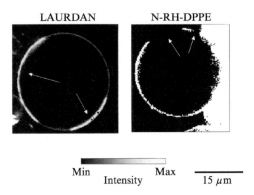

Min   Max   ───────
    Intensity   15 μm

Fig. 4. Two-photon excitation fluorescence intensity images of LAURDAN- and N-Rh-DPPE–labeled GUVs composed of DPPE–DPPC (7:3, mol/mol) at the phase coexistence temperature region. The images have been taken at the equatorial region of the GUV. The open arrows indicate gel phase regions (see text). Note that the gel domains span the lipid bilayer. The probe concentration was below 0.25 mol%.

Photoselection Effect and LAURDAN Generalized Polarization Function

In addition, the parallel location of the electronic transition dipole of LAURDAN with respect to the lipid chains in the bilayer offers an important advantage to ascertain lipid phase coexistence from the fluorescence images, using the photoselection effect.[43–45,48,89] The photoselection effect is dictated by the fact that only those fluorophores that possess the electronic transition dipole aligned parallel, or nearly so, to the plane of polarization of the excitation light become excited. Using circular polarized excitation light (on the $x–y$ plane) and observing the polar region of a LAURDAN-labeled GUV displaying phase coexistence, only fluorescence emanating from the fluid part of the bilayer is expected because the relatively low lipid order allows a component of the probe's electronic transition dipole to parallel the polarization plane of the excitation light. This last fact is well represented in Fig. 5, which shows fluorescence images of LAURDAN-labeled GUVs composed of DLPC–DPPC and DPPE–DPPC at the phase coexistence temperature region; the images were taken at the top part of the vesicle. The line-shaped domains in DLPC–DPPC and the leaf-shaped domains in DPPC–DPPE are gel-phase domains as judged from the LAURDAN fluorescence image (there is a strong photoselection effect in these regions).[44,45] Taking into account this last fact, N-Rh-DPPE partition is favored to the gel phase in DLPC–DPPC and to the fluid phase

[89] T. Parasassi, E. Gratton, W. Yu, P. Wilson, and M. Levi, *Biophys. J.* **72**, 2413 (1997).

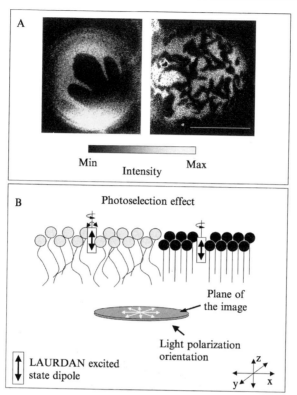

FIG. 5. (A) Two-photon excitation fluorescence intensity images of LAURDAN-labeled GUVs composed of DPPE:DPPC (7:3, mol/mol) (*left*) and DLPC:DPPC (1:1, mol/mol) (*right*). (B) Sketch of LAURDAN orientations on the lipid bilayer with respect to the polarization plane of the excitation light. The lipid molecules in the fluid and gel phases were colored gray and black, respectively, as observed in the two-photon excitation images (A). Note that the wobbling movement of LAURDAN due to the loosely packed lipid organization in the fluid phase allows excitation of a component of the probe's transition dipole. This last phenomenon is not observed in the gel phase (photoselection effect, see text).

in DPPC–DPPE (compare Figs. 3 and 5).[44,45] It is important to remark that the photoselection effect in the case of LAURDAN is evident only at the polar region of the vesicle, where the probe molecules are located parallel to the lipid chains along the z axis (the polarization plane of the excitation light is on the x–y plane).[48]

The fact that LAURDAN has different emission spectra between the coexisting lipid phases is well exploited by calculating LAURDAN generalized polarization (GP) function images.[43–48,89] The GP is a useful relationship between the emission intensities obtained from the blue and red sides of the emission spectrum and contains information about solvent dipolar relaxation processes that occur during the time that LAURDAN is in the excited state, and is related to water penetration in phospholipid interfaces and therefore is related to the phase state of the lipid membrane.[84,85] The bluer the emission spectrum the higher the GP value. Calculations of LAURDAN GP images are confined to images obtained at the equatorial region of vesicles to avoid artifacts introduced by the photoselection effect, that is, the same excitation efficiency is expected for LAURDAN molecules at the equatorial region of GUVs displaying phase coexistence when using circular polarized excitation light because all LAURDAN molecules have the transition dipole aligned parallel to the polarization plane of the excitation light.[48] This last effect is well represented in Fig. 6, which shows a coexistence between two different fluid phases [dioleoylphosphatidylcholine (DOPC)–cholesterol–sphingomyetin (1:1:1, mol/mol)].[47] Note that the photoselection effect produces a progressive decrease in LAURDAN GP value in each different lipid phase, being stronger in the more ordered phase as the pole of the vesicle is approached

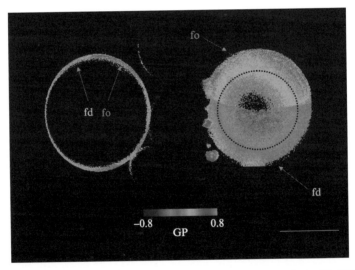

FIG. 6. Two-photon excitation GP images (false color representation) of LAURDAN-labeled GUVs composed DOPC–cholesterol–sphingomyelin (1:1:1, mol/mol) displaying fluid-ordered (fo)/fluid-disordered (fd) phase coexistence (false color representation). GP images obtained at the equatorial region of the GUV, where the photoselection effect is prevented (*left*) and at the polar region of the GUV, where the photoselection operates (*right*). Note the influence of the photoselection effect on the LAURDAN calculated values (dotted black circle) in the image obtained at the polar region of the GUV. Scale bar: 20 nm. (See Color Insert.)

(Fig. 6, right, dotted circle). In addition, Fig. 6 (left) shows a GP image taken at the equatorial region of the GUV, where the photoselection effect is prevented.

LAURDAN GP images provide information about phase coexistence, even though the lipid domains are smaller than the microscope resolution, by using linearly polarized excitation light. The reader is encouraged to explore Refs. 43 and 89, which confirm the advantages of using this experimental approach in different lipid systems. To summarize this section, when the lipid domains are bigger than the microscope resolution ($\sim$0.3 μm), exciting LAURDAN with circularly polarized light allows one to extract information about the lipid domain phase state and shape by (1) the orientation of the lipids in a particular domain (polar region, photoselection effect) and (2) the extent of water dipolar relaxation processes in the different lipid domains (equatorial region, GP).[48] LAURDAN and also PRODAN[44,48,90] are unique fluorescent probes that enable the extraction of simultaneous information about lipid domain phase state and shape from fluorescence images.

Direct Visualization of Lipid Domains

The quality and novelty of the information extracted using the experimental approach described previously is remarkable. For example, one of the most relevant observations made through the direct visualization of domain coexistence in the many different mixtures (artificial or natural lipid mixtures) studies is that the lipid domains span the lipid bilayer.[44,45,47,48,55,74] Two important observations, independent of the nature of the fluorescent probe or lipid mixture studied, support the last conclusion: (1) in observing Fig. 4, for example, it is clear that both N-Rh-DPPE and LAURDAN—which are localized in both leaflets of the bilayer—show the spanning of the lipid domain at the equatorial region of the vesicle because N-Rh-DPPE is excluded from the gel phase and LAURDAN displays different emission spectra from each lipid phase; and (2) the domain spanning the lipid bilayer is extracted from fluorescence images taken at the polar region of the GUVs. Namely, the lipid domain shape does not change, regardless of the fluorescent probe used, at constant temperature for a long period of time. This last fact is well illustrated in Fig. 7, in which fluorescence images obtained at the polar region of a giant multilamellar vesicle (GMV, two bilayers in this particular case) observed as a function of time are compared with those obtained in GUVs at constant

[90] E. K. Krasnowska, L. A. Bagatolli, E. Gratton, and T. Parasassi, *Biochim. Biophys. Acta* **1511,** 330 (2001).

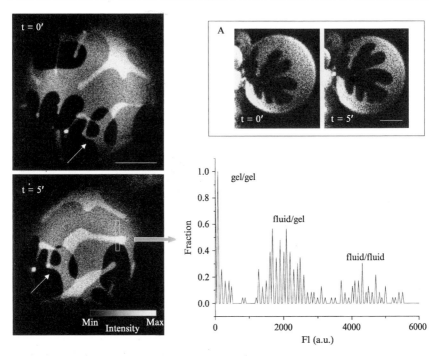

FIG. 7. Two-photon excitation fluorescence intensity images of *N*-Rh-DPPE–labeled giant multilamellar vesicle (GMV, two bilayers in this case) composed of DPPE–DPPC (7:3, mol/mol) at the phase coexistence temperature region. The images have been taken at the top part of the GMV. The open arrow in the GMV fluorescence image indicates independent movement of the two bilayers, showing the particular pattern of uncoupled lipid layers. This particular pattern is not observed in GUVs (*inset A*), showing that the gel domains in both leaflets of the bilayer are coupled. The intensity histogram obtained from the boxed region (bottom GMV image) shows three different situations: (1) overlap of two gel bilayers (black), (2) overlap of gel and fluid bilayers (gray), and (3) overlap of two fluid bilayers (intense gray). Note that the mean intensity value in the intense gray area duplicates that obtained in the gray area (see text). Scale bar: 10 μm.

temperature. As observed in Fig. 7, the fluorescence intensity where two fluid bilayer regions are superimposed is twice the value of a fluid bilayer region superimposed with a gel bilayer region. In addition, a lack of fluorescence intensity is observed only when two gel bilayer regions are superimposed. GMV images are clear examples of uncoupled lipid layers. This last is not observed in GUVs[44,45,47,48,55,74] (see inset A in Fig. 7), supporting the fact that the domain spans the lipid bilayer. Domains spanning the membrane confirm the existence of epitactic coupling in free-standing bilayers. Direct epitactic coupling has also been demonstrated in studies of supported membranes by Merkel *et al.*,[91] who showed that laterally ordered domains in one monolayer induce solidification of domains of identical topology in the juxtaposed monolayer transferred from the fluid state. Because this phenomenon is independent of the composition[44,45,47,48,55,74] it is possible that this last feature could provide a mechanism for how events that occur in the outer monolayer of a biological membrane are coupled to cytoplasmic components of signal transduction pathways—a question that has eluded simple answers.[92]

Lipid Domain Shape Information

The phase coexistence observed at the level of single vesicles can be divided into three different groups: gel/fluid-phase coexistence,[44,45,74] fluid-ordered/fluid-disordered phase coexistence,[47,93] and a peculiar phase coexistence observed in binary mixtures composed of bipolar lipids from thermoacidophilic archaebacteria, for example, *Sulfolobus acidocaldarius*.[46]

The main differences observed between gel/fluid and fluid-ordered/ fluid-disordered phase coexistence is related to the domain shape and the behavior of LAURDAN[48] (Fig. 8). Two general trends related to the shape of lipid domains and LAURDAN fluorescence properties are observed in samples displaying gel/fluid-phase coexistence, namely, gel domains with irregular shape and a lack of fluorescence intensity due to strong effect of photoselection (Fig. 8A). In addition to well-characterized phospholipid binary mixtures, which display gel/fluid-phase coexistence,[44,45] natural samples such as bovine lipid extract surfactant[74] or brush border membrane lipid extracts after cholesterol extraction[47] can be included in this group even though these last samples display a heterogeneous lipid composition (see Fig. 8A). Also, important subtle differences among all these mixtures can be extracted on the basis of LAURDAN GP images, as described in the next section.

On the other hand, a common characteristic of cholesterol-containing mixtures (natural and artificial lipid mixtures)[47] is the circular shape of the domain at the phase coexistence temperature region, something that is not observed in samples displaying gel/fluid-phase coexistence (compare Fig. 8A and B). When fluid domains are embedded in a fluid environment, circular domains will form because both phases are isotropic and the line energy (tension), which is associated with the rim of two demixing phases, is minimized by optimizing the area-to-perimeter ratio.[47] Moreover, the strong photoselection effect observed in the gel domains is not observed in the fluid-ordered domains in GUVs composed of

[91] R. Merkel, E. Sackmann, and E. Evans, *J. Phys. France* **50,** 1535 (1989).
[92] D. A. Brown and E. London, *Annu. Rev. Cell Dev. Biol.* **14,** 111 (1998).
[93] S. L. Veatch, and S. L. Keller, *Phys. Rev. Lett.* **89,** 268101 (2002).

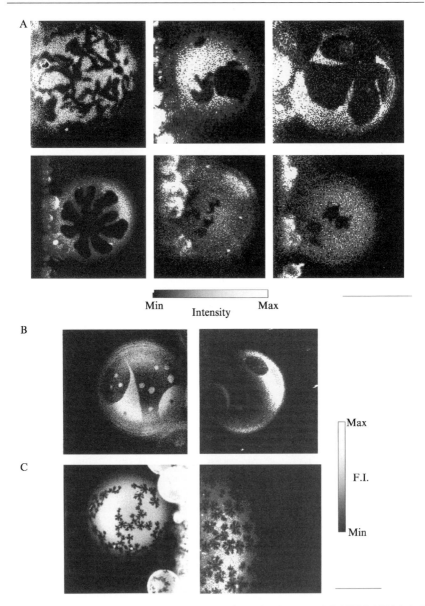

Fig. 8. Two-photon excitation fluorescence intensity images of LAURDAN-labeled GUVs at the phase coexistence temperature region. The images have been taken at the equatorial region of the GUV. (A) Mixtures displaying gel/fluid phase coexistence: DLPC/ DPPC, DLPC/DSPC, and DLPC/DAPC (1:1, mol:mol) (*top*, from left to right); DMPC/ DMPE (1:1, mol/mol), cholesterol-depleted brush border membrane lipid extract and bovine

DOPC–cholesterol–sphingomyelin mixtures, indicating that the fluorescent probe lies in an ordered but fluid environment (see Fig. 8B, left image).

A particular phase coexistence was observed in a binary mixture composed of bipolar lipids extracted from *S. acidocaldarius* [polar lipid fraction E (PLFE) lipids].[46] The reason this particular phase coexistence is classified in a third group is mainly because the PLFE GUV membrane is a monolayer. In this case lipid domains having a "snowflake" shape are detected (see Fig. 8C). The LAURDAN location in these membranes is also peculiar, that is, LAURDAN has the transition dipole aligned parallel to the vesicle surface and the chromopore moiety lies at the level of the polar head groups.[46] Following the LAURDAN GP a phase transition related to the reorientation of the polar head groups was detected at 50° in this mixture. However, the appearance of phase coexistence was detected around 20°, with LAURDAN being excluded from the snowflake domains. These results suggest the possibility of lipid lateral segregation in the presence of rare lipid species and open new questions about the biological relevance of such phenomena in complex lipid systems.

LAURDAN GP Can Be Related to Compositional Differences between Coexisting Lipid Domains

In addition to domain shape information and qualitatively inferences about the lipid domain phase state, using the photoselection effect, LAURDAN GP images can offer important subtle information about particular characteristics of the coexisting phases. An interesting example is that reported for binary phospholipid mixtures that contain DLPC.[45] In this experiment the hydrophobic mismatch of the phospholipid binary mixtures was increased by using DPPC, distearsylphosphatidylcholine (DSPC), and diarachidonylphosphatidylcholine (DAPC) as the second lipid component. The first comparison among these mixtures reveals that the lipid domain shape changes as the hydrophobic mismatch of the binary mixtures increases. This last conclusion is accompanied by additional qualitative information extracted from the LAURDAN intensity images taken at the polar region of the GUVs (where the photoselection effect operates), that is, gel and fluid domains coexist in the lipid bilayer. An important question to be answered is whether the general characteristics of the phase coexistence picture obtained present any correspondence among the different lipid mixtures.

---

lipid extract surfactant (*bottom*, from left to right). (B) Mixtures that display fluid-ordered/fluid-disordered phase coexistence: DOPC–cholesterol–sphingomyelin (1:1:1, mol/mol) with 1 mol% of ganglioside G_{M1} (*left*) and brush border membrane lipid extract (*right*). (C) PLFE lipids displaying phase coexistence. The probe concentration was below 0.25 mol%. Scale bar: 30 micrometers.

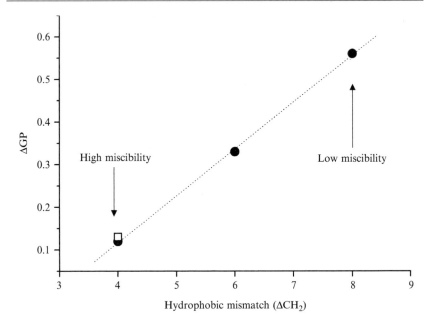

Fɪɢ. 9. LAURDAN GP differences between the fluid- and gel-phase domains (ΔGP) plotted versus the hydrophobic mismatch between the components of the binary mixtures. (●) DLPC-containing GUVs [DLPC–DPPC, DLPC–DSPC, and DLPC–DAPC (1:1, mol/mol), hydrophobic mismatch equal to 4, 6, and 8 respectively]; (□) DMPC–DSPC mixture (1:1, mol/mol; hydrophobic mismatch equal to 4). The ΔGP between the gel and fluid components for DMPE–DMPC and DPPE–DPPC (different polar head group/same lipid chain length) was similar to that observed for DLPC–DAPC, showing similar leaf shape (see Fig. 8A).

Subtle differences in phase coexistence among the different mixtures can be obtained from GP images. For example, differences in GP values between the gel and fluid regions of the lipid bilayer (ΔGP) plotted versus the hydrophobic mismatch (ΔCH$_2$) of DLPC-containing binary phospholipid mixtures shows a linear relationship, that is, the difference in the extent of water dipolar relaxation between the gel and fluid domains increases as the hydrophobic mismatch increases (see Fig. 9).[45] This last result is explained by taking into account the progressive increment in the compositional and energetic differences between the gel and fluid domains as the miscibility between the binary phospholipid mixture components decreases.[48] As shown in Fig. 9, similar ΔGP values are obtained for dimyristoylphosphatdylcholine (DMPC)–DSPC and DLPC–DPPC, in agreement with the fact that these mixtures have the same hydrophobic mismatch. Other samples that display low miscibility, such as DPPC–DPPE and DMPC–DMPE, display a ΔGP similar to that observed for

DLPC–DAPC. Interestingly, these low-miscibility phospholipid mixtures also present a similar shape for the gel domain (leaf shape).[45]

Conclusions

A new experimental approach to study lipid–lipid interactions has been presented in this chapter. The enormous advantages in using the fluorescent probe LAURDAN as a tool to study lipid phase coexistence is noteworthy. Important novel information such as lipid domains spanning the lipid bilayer, correlation of domain shape with lipid mixture composition, and comparisons of lipid domain compositional and energetic differences among different phospholipid mixtures is extracted directly from fluorescence images. The experiments at the level of single vesicles undoubtedly open new ways to study the thermotropic behavior of lipid bilayers, including the possibility to explore lipid–protein interactions.[51,94]

Acknowledgments

This research was supported by Fundacion Antorchas (Argentina), Beca de Investigación Carrillo-Oñativia, Ministerio de Salud de la Nación (Argentina), and RR03155 (which supports the Laboratory for Fluorescence Dynamics). The author thanks Dr. E. Gratton for stimulating opinions about the experimental results and for support during the author's postdoctoral research in his laboratory, and Dr. D. M. Jameson for critical reading of the manuscript. MEMPHYS-Center for Biomembrane Physics is supported by the Danish National Research Foundation.

[94] L. A. Bagatolli, S. Sanchez, T. Hazlett, and E. Gratton, *Methods Enzymol.* **360**, 481 (2003).

[16] Determination of Liposome Surface Dielectric Constant and Hydrophobicity

By S. Ohki and K. Arnold

Introduction

Phospholipids of biological origin usually form stable bilayer membranes in aqueous solutions, exposing their polar groups to the surrounding aqueous phases and enclosing their nonpolar hydrocarbon tails in the interior of the membrane.[1] When such bilayers are in an aqueous solution, they

[1] R. Harrison and G. G. Lunt, "Biological Membranes." Blackie, London, 1980.

form the lamellar phase. However, depending on changes in the surface nature of such lipid bilayers caused by structural alterations of lipid molecules and/or changes in the environmental conditions, (e.g., ionic concentrations or temperature), they can lose their stability as bilayers and take on various shapes and surfaces.[2] The surface nature of the lipid bilayer is also an important factor in discussing the interaction of two bilayers, in such events as adhesion and fusion. For example, as long as the distance of two bilayers is greater than $\sim$20 Å, the physical behavior of the two interacting bilayers can be explained by the so-called DLVO theory,[3] which does not include the nature of the membrane surfaces. However, because most lipid bilayer surfaces are strongly hydrophilic, water molecules bind to their surfaces and form the structured water. As soon as the intermembrane distance becomes shorter than $\sim$20 Å, hydration interaction forces[4,5] show their effect on the interacting membranes. A hydration interaction term then must be included in the DLVO theory, yielding the "modified DLVO theory,"[6] to explain satisfactorily intermembrane interaction.

The hydration of the membrane surface is strongly related to its hydrophobicity. Although the term "hydrophobicity" in general has wider implications, we use the term "surface hydrophobicity" here in a limited definition as the degree of hydrocarbon-like nature of the surface regions. Thus, hydrophobicity can be defined as the degree of excess free energy of the interface facing the aqueous phase, or the degree of increase in interfacial tension against the aqueous phase. The polarity of the membrane surface is somewhat related to the interfacial tension as well as its dielectric properties. These quantities seem to have some interrelation to each other. In this chapter, the measurements of such quantities as surface dielectric constant, surface hydrophobicity, and interfacial tension of liposomes are described and some discussion regarding the interrelation of these quantities is made in relation to some biologically relevant events, such as membrane adhesion and fusion.

[2] M. D. Houslay and K. K. Stanley, "Dynamics of Biological Membranes." John Wiley & Sons, New York, 1984.

[3] J. Th. G. Overbeek, in "Colloid Science" (H. Kruyt, ed.), Vol. 1. Elsevier, Amsterdam, 1952.

[4] D. M. LeNeveu, P. Rand, and V. A. Parsegian, Nature 259, 601 (1976).

[5] P. Rand and V. A. Parsegian, in "The Structure of Biological Membranes" (P. Yeagle, ed.), p. 251. CRC Press, Boca Raton, FL, 1991.

[6] S. Ohki and H. Ohshima, Colloids Surfaces B Biointerfaces 14, 27 (1999).

Surface Dielectric Constant of Lipid Vesicles

Measurements of Dielectric Constant of Liposome Surfaces by Fluorescence Spectroscopy

The surface dielectric constant of lipid membranes should vary with the distance from the membrane surface. The dielectric constant of the interior of lipid membranes would be about 2.0 and, as the measurement position moves from the interior of the membrane to the exterior of an aqueous phase along a line normal to the membrane surface, the dielectric constant would vary from 2 to 80 (Fig. 1). Therefore, the values of the surface dielectric constant of the membrane would be different depending on where the dielectric constant is determined. In this chapter, a method to measure the dielectric constant of the lipid membrane surface is described that measures the properties of an environment-sensitive fluorophore attached to the lipid polar group. The fluorophore is considered to be located in the region of the glycerol backbone according to a previous study.[7]

Materials and Experimental Procedures

Materials. Phospholipids [phosphatidylcholine (PC), phosphatidyletha-nolamine (PE), phosphatidic acid (PA), and phosphatidylserine (PS)] and fluorescently labeled phospholipid (dansylphosphatidylethanolamine, DPE) can be obtained from Avanti Polar Lipids (Alabaster, AL). Organic solvents (methanol, ethanol, 1-propanol, chloroform, 1-octanol, *n*-butylamine, and *n*-decane) should be of reagent grade and are obtained from J. T. Baker Chemical (Phillipsburg, NJ). Octylamine is obtained from Aldrich Chemical (Milwaukee, WI). HEPES (*N*-2-hydroxyethylpipera-zine- *N'*-2-ethanesulfonic acid) can be obtained from Calbiochem (San Diego, CA). Trivalent cation salts of $TbCl_3 \cdot 6H_2O$ and $LaCl_3 \cdot 7H_2O$ of the highest purity can be obtained from Aldrich Chemical. Poly(ethylene glycol) (MW 6000) can be purchased from Fluka Chemika Biochemika (Buchs, Switzerland). All other chemicals are of reagent grade and can be obtained from Baker Chemical.

Preparation of Samples

Fluorescent Probe Solutions. Dansylphosphatidylethanolamine (DPE) is suspended in various organic solvents (*n*-decane, octylamine, *n*-butyl-amine, 1-octanol, 1-propanol, ethanol, or methanol) at 15 μg/ml and the suspension is vortexed for 15 min.[8] Because the dansyl fluorophore is sensitive

---

[7] A. S. Waggoner and L. Stryer, *Proc. Natl. Acad. Sci. USA* **67,** 579 (1970).

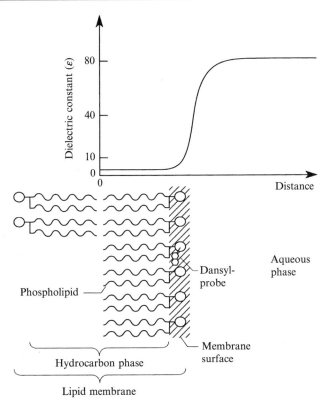

FIG. 1. Schematic sketch of the dielectric constant from the center of a lipid bilayer to an aqueous solution.

to light, the probe solutions should be protected from light and kept in a refrigerator until use.

Small Lipid Vesicles Bearing Fluorophore. Small unilamellar lipid vesicles (SUVs) are prepared by hydrating 5 mg of phosphatidylcholine (PC) or phosphatidylserine (PS) and DPE at a molar ratio of lipid to DPE of 200–300 in 1 ml of salt solution, for example, 100 mM NaCl, 10 mM HEPES, pH 7.0 (NHB), vortexing for 10 min, and sonicating for 30 min in a bath-type sonicator (model G11; Laboratory Supplies, Hicksville, NY) until the lipid suspension becomes clear. With these procedures, the average size of small liposomes should be approximately 30–40 nm in diameter. If the suspension does not become clear, the suspension should be

---

[8] S. Ohki and K. Arnold, *J. Membr. Biol.* **114,** 196 (1990).

centrifuged at 100,000 g for 1 h and the supernatant, which contains mostly SUVs, should be collected.

Large Lipid Vesicles Bearing Fluorescence Probe. Five milligrams of phospholipid (PC or PS) and DPE at the same molar ratio (lipid–DPE, 200–300) as the SUV vesicles, is hydrated with 1 ml of NHB and the suspension is vortexed for 10 min to form multilamellar vesicles (MLVs). Large unilamellar vesicles (LUVs) are prepared by passing the MLV suspension through polycarbonate filters with a pore diameter of 0.1 μm (Nuclepore, Pleasanton, CA) about 20 times by the use of two syringes[9] (Hamilton, Reno, NV). Another extrusion method is by means of the Extruder[10] (Lipex Biomembranes, Vancouver, BC, Canada), through which MLVs are pushed by high pressure. The filtered solution is a suspension of LUVs with an average diameter of 110–130 nm, as determined by a submicron particle analyzer utilizing dynamic light scattering (e.g., model N4; Coulter, Hialeah, FL).

Measurements of Dansyl Fluorescence. The emission maximum of dansyl fluorescence in methanol is approximately 515 nm by exciting at 350 nm. The emission spectrum can be taken in the range of 450–550 nm. Such measurements can be made with a spectrofluorometer (e.g., LS-5; Perkin Elmer, Norwalk, CT).

Calibration of Dielectric Constants of Environment of DPE in Various Solvents. The fluorescence of the dansyl fluorophore depends strongly on its environmental polarity; the lower the dielectric constant of the environment, the shorter the wavelength of the maximum of the emission spectrum becomes (Stokes shift).[11] To establish a baseline to estimate the dielectric constant of the environment for the fluorophore, a relationship between the known dielectric constants of various organic solvents and the wavelength at the maximum in the fluorescence emission spectrum of DPE in each organic solvent is obtained. For this purpose, the emission spectra (400–600 nm) of DPE in various organic solvents (e.g., methanol, ethanol, 1-propanol, 1-octanol, *n*-butylamine, octylamine, and *n*-decane) may be measured by exciting the fluorophore at 340 ± 5 nm.[8,12] According to the theory,[13] the wavelength at the spectrum maximum of the fluorophore is related to the dielectric constant and the refractive index of the medium according to the following formula:

[9] R. C. MacDonald, R. I. MacDonald, B. P. M. Menco, K. Takeshita, N. K. Subbarao, and L. Hu, *Biochim. Biophys. Acta* **1061**, 297 (1991).

[10] L. D. Mayer, M. J. Hope, and P. R. Cullis, *Biochim. Biophys. Acta* **858**, 161 (1986).

[11] J. R. Lakowicz, "Principles of Fluorescence Spectroscopy." Plenum Press, New York, 1986.

[12] Y. Kimura and A. Ikegami, *J. Membr. Biol.* **85**, 225 (1985).

[13] E. von Lippert, *Z. Electrochem.* **61**, 962 (1957).

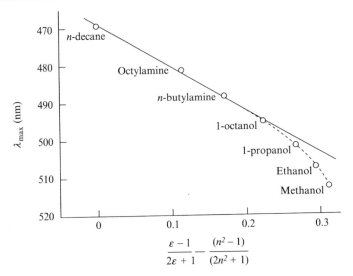

FIG. 2. The relationship of dielectric constant and λ_{max} of DPE fluorescence determined in various solvents: methanol ($\varepsilon = 32.6$, $n = 1.329$), ethanol (24.3, 1.36), 1-propanol (20.1, 1.385), 1-octanol (10.3, 1.43), n-butylamine (5.4, 1.403), octylamine (3.4, 1.429), and n-decane (1.991, 1.410). The values of dielectric constants and refractive indices are taken from Ref. 35. Figure modified from Ref. 8.

$$1/\lambda_{max} = K\left[\frac{(\varepsilon - 1)}{(2\varepsilon + 1)} - \frac{(n^2 - 1)}{(2n^2 + 1)}\right] \tag{1}$$

where λ_{max} is the wavelength of the emission spectrum maximum, K is a constant, and ε and n are the dielectric constant and the refractive index of the medium, respectively. The experimental relationship between the wavelength, λ_{max}, and the dielectric properties of the media are shown in Fig. 2.[8] Another plot of the relation between dielectric constant and the maximum wavelength of the observed emission spectra is shown in Fig. 3.[14]

Surface Dielectric Constant of PS Vesicles with Various Cations

Emission Spectra. The emission spectrum of DPE incorporated in liposomes suspended in various salt solutions is measured as a function of salt concentrations. As the fusogenic ion (polyvalent cations, and also hydrogen ion) concentration is increased, the maximum position of the emission spectrum shifts toward the higher frequency (shorter wavelength)

[14] S. Ohki and O. Zschörnig, *Chem. Phys. Lipids* **65**, 193 (1993).
[35] R. C. Weast, ed., "Handbook of Chemistry and Physics," 49th Ed. Chemical Rubber Company, Cleveland, OH, 1948.

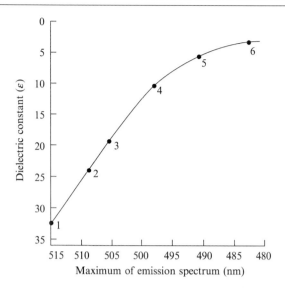

FIG. 3. The relationship between the maximum position of fluorescence spectra of DPE in various organic media and their dielectric constants. (1) Methanol; (2) ethanol; (3) 1-propanol; (4) 1-octanol; (5) *n*-butanol; (6) octylamine. From Ref. 14.

of the light and the fluorescence intensity also increases. These changes (the blue shift of the emission spectrum and an increase in fluorescence intensity) indicate that the local environment around the fluorophore is altered to a further nonpolar; lower dielectric medium. Because the fluorophore (dansyl group) of DPE is to detect the dielectric medium of the glycerol backbone region of the lipid bilayer,[7] the observed changes in fluorescence signal indicate that the polar region (particularly, the glycerol backbone region) of the bilayer becomes lower in dielectric nature (or more nonpolar, or hydrophobic in nature), on interaction with fusogenic ions.

Surface Dielectric Constants of Liposomes Determined from Emission Spectra. With a calibration curve (Fig. 3) of the relationship between λ_{max} and the known dielectric media, and the observed wavelength of the emission spectra maxima of DPE as a function of salt concentration, we can obtain the dielectric constant of the lipid polar (glycerol backbone) region for lipid vesicles. The surface dielectric constants of small unilamellar PS vesicles are shown in Fig. 4.[8] The surface dielectric constant of the PS membrane is about 30 in 0.1 M NaCl, and the values of the dielectric constant are reduced as the concentration of fusogenic ions increases; for example, for 1 mM Ca^{2+} in 0.1 M NaCl, the surface dielectric constant of the PS small vesicles is approximately 8. A further increase in the

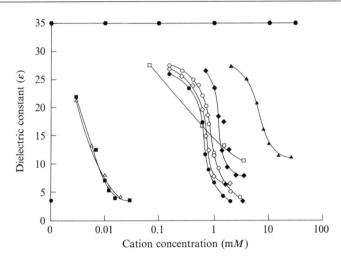

FIG. 4. Surface dielectric constants of the environment for DPE in PS vesicles suspended in 0.1 M NaCl buffer containing various fusogenic cation concentrations ($\triangle$, La^{3+}; $\blacksquare$, Tb^{3+}; $\bullet$, Mn^{2+}; $\diamond$, Ba^{2+}; $\bigcirc$, Ca^{2+}; $\blacklozenge$, Sr^{2+}; $\square$, H^{+}; $\blacktriangle$, Mg^{2+}). ($\bullet$) The control experiment using PC vesicles with respect to Ca^{2+}. From Ref. 8.

calcium ion concentration can reduce the surface dielectric constant to as low as 3.

It is well known that poly(ethylene glycol) dehydrates the membrane surface at high concentrations by pressing two membranes together.[15–17] It is, therefore, expected that the surface dielectric constant of liposomes would become low at high concentrations of PEG in a liposome suspension and would decrease as the concentration of poly(ethylene glycol) increases. Such measurements can be demonstrated by measuring the emission spectra shift of DPE incorporated into lipid vesicles in an aqueous medium containing various amounts of poly(ethylene glycol) [e.g., 5, 10, 15, 20, and 30% (w/w)]. The surface dielectric constants measured for lipid vesicles in the presence of various concentrations of PEG are shown in Fig. 5 for liposomes made of PC, PS, and PS–PE (1:1) in NHB. The surface dielectric constant of the vesicles is indeed reduced as the concentration of PEG increases; PEG causes dehydration of the membrane surfaces. This is another confirmation that the PC membrane surface is most hydrophilic

[15] K. Arnold, L. Pratsch, and K. Gawrisch, *Biochim. Biophys. Acta* **728,** 121 (1983).
[16] K. Arnold, O. Zschörnig, D. Barthel, and W. Herold, *Biochim. Biophys. Acta* **1022,** 303 (1990).
[17] S. W. Burgess, T. J. McIntosh, and B. R. Lentz, *Biochemistry* **31,** 2653 (1992).

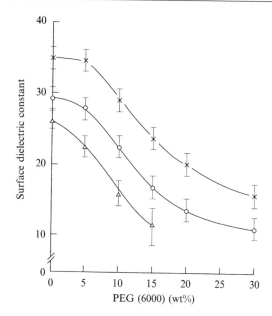

FIG. 5. Surface dielectric constants of the environment for DPE in PC, PS, and PS–PE (1:1) vesicles suspended in 0.1 M NaCl buffer as a function of PEG (6000) concentrations. ×, PC vesicles; ○, PS vesicles; △, PS–PE (1:1) vesicles. From Ref. 8 and unpublished data for PS–PE (1:1).

and the PE membrane is least hydrophilic among these three lipid membranes used, and the order of the magnitude of the surface dielectric constants of the liposomes of PC, PS, and PE–PS (1:1) is PC > PS > PS–PE (approximately 34, 30, and 27 in 0.1 M NaCl buffer, pH 7.0).

Similar measurements can be performed for LUVs as a function of polycation or extrinsic protein concentrations. Some experimental results[18,19] are given in Table I, together with those for SUVs.[8,14,20] In 0.1 M NaCl, the surface dielectric constant for LUVs is lower than that for SUVs. However, at elevated concentrations of fusogenic cations or poly(ethylene glycol), the dielectric constant for LUVs does not become as low as that for SUVs (Table I).

[18] G. Köhler, U. Hering, O. Zschörnig, and K. Arnold, *Biochemistry* **36**, 8189 (1997).
[19] H. Binder, G. Köhler, K. Arnold, and O. Zschörnig, *Phys. Chem. Chem. Phys.* **2**, 4615 (2000).

TABLE I
SURFACE DIELECTRIC CONSTANTS FOR VARIOUS LIPID VESICLES AT DIFFERENT IONIC CONDITIONS

Ions	Conc. (mM)	SUV				LUV[a]	
		PC[b]	PS[b]	PA[c]	PI[d]	PS	PA
NaCl	100	34	30	32	37	23	24
+Ca^{2+}	0.4	34	25	13	-	21	22
	1.0	34	12	5	19	19	20.5
	2.0	34	5	-	16	14	13
	5.0	34	-	-	14	10	9
	10	34	-	-	12.5	7	6
+Mg^{2+}	0.4	34	-	14			
	1.0	34	-	9			
	2.0	34	27	-			
	5.0	34	16	-			
	10	34	11	-			

[a] Data from Köhler et al. (1997); Binder et al. (2000).
[b] Data from Ohki and Arnold (1990).
[c] Data from Ohki and Zschörnig (1993).
[d] Data from Müller et al. (2003).

Determination of Surface Hydrohobicity of Liposome Membranes

The surface hydrophobicity of liposome membranes may be expressed as an increase in the excess free energy at the interface of the membrane and aqueous phases. Therefore, it is expressed in terms of the interfacial tension, which is not the film tension. Although the interfacial tension of lipid vesicle membranes is difficult to measure, two types of experiments provide useful information in this regard: one of them is a comparative experiment measuring the interfacial tension of a lipid monolayer formed at the oil–water interface,[21] and the other is a direct measurement of the interfacial tension of the liposome membrane surface interacting with an aqueous solution by the contact angle method, utilizing Young's equation[22] and a theory of the surface energy, a method developed by van Oss and others.[23] Nevertheless, accurate measurements of the interfacial energy for lipid membranes with the latter method still encounter some technical difficulties.

[20] M. Müller, O. Zschörnig, S. Ohki and K. Arnold, *J. Membrane Biol.* **192**, 33 (2003).
[21] J. T. Davies and E. K. Redial, "Interfacial Phenomena." Academic Press, New York, 1961.
[22] R. E. Johnson and R. Dettre, in "Surface and Colloid Science" (E. Matijevic, ed.), Vol. 1, p. 85. Wiley-Interscience, New York, 1969.
[23] C. J. van Oss, "Interfacial Forces in Aqueous Media." Marcel Dekker, New York, 1994.

Lipid Monolayer Studies

Membrane surface hydrophobicity of lipid vesicles can be deduced from the measurements of the interfacial tension of lipid monolayers formed at the oil–water interface.

Methods and Procedures

See Refs. 24 and 25 for more details.

Chemicals. Phospholipids can be obtained from Avanti Polar Lipids, as mentioned in the previous section. Hexane and methanol of the Baker Instra-Analyzed grade, or of the best quality available, are used. Salts, for example, NaCl, are of reagent grade and should be roasted at 600–700° for 1 h before use in these experiments in order to remove any organic impurities. Water to be used is distilled in a glass apparatus, including the processes of deionizing and alkaline permanganate distilling, so that any contaminants of organic substances are absolutely minimal. To remove surface-active impurities, hexadecane [purum > 99% (GC) of Fluka or better grade] is further purified by washing with water vigorously and leaving the suspension for 1 h to phase separate into two phases in a separation funnel. Then, only the upper oil phase is collected, particularly avoiding the interface region by use of a separation funnel. This procedure should be repeated at least three times.

Monolayer-Forming Solution. A small amount (~ 0.5 mg) of phospholipid (e.g., PS or PC) is dissolved in 1.2 ml of hexane–methanol (1:1, v/v) to make an approximately 0.5×10^{-4} M lipid solution.

Measurements of Surface Tension of Lipid Monolayers Formed at Air–Water Interface and Oil–Water Interface

Surface Tension Measurements of Monolayers Formed at Air–Water Interface. An aqueous solution (e.g., 0.1 M NaCl buffer, NHB) is placed in a circular glass trough, 6 or 9 cm in diameter.[25] The tip of a Wilhelmy plate made of 18 mm (w) × 8 mm (h) × 0.16 mm (thickness) microscope coverslip fixed (via heat) with a thin platinum wire, which is hanged on one of the arms of an electronic balance, is immersed vertically into the aqueous solution. This immersion results in a downward force on the glass plate due to the wetting of the glass with water. This force is measured by use of an electronic balance (e.g., Cahn Microbalance, Model 28). The force is the

[24] C. B. Ohki, S. Ohki, and N. Düzgüneş, "Colloid and Interface Science" (M. Kerker, ed.), Vol. V, p. 271. Academic Press, New York, 1976.
[25] S. Ohki, *Biochim. Biophys. Acta* **689**, 1 (1982).

product of the surface tension of water and the perimeter of the glass plate, which is 72 dyn/cm (water surface tension) × 2 (18 + 0.16) mm ≅ 260 dyn. Conversely, by measuring the force, the surface tension of other aqueous solutions can be obtained. The surface tension of 0.1 M NaCl is approximately the same as that of water. An aliquot of the monolayer forming solution is placed onto the aqueous solution surface. As the lipid monolayer forms, the surface tension of the aqueous phase covered by the monolayer decreases. When monolayer formation is completed, a stable tension will be observed, usually within a minute. If a stable surface tension is not obtained, either the lipid samples or aqueous solution, or both, may contain impurities. For given amounts of lipids applied onto the surface, the areas per lipid molecule are calculated. The measured surface tension results are usually shown by the pressure, π, of the film, which is expressed by $\pi = \gamma_w - \gamma$, where γ_w and γ are the surface tensions of water and the monolayer, respectively. The measured pressure for a PS monolayer formed at the air–water interface with respect to area (A) per molecule is shown in Fig. 6.[24,25]

Formation of Lipid Monolayer at the Oil–Water Interface and Measurements of Interfacial Tension. A circular glass trough with a diameter of 6 or 9 cm and a depth of 5 cm can be used. The upper portion (about

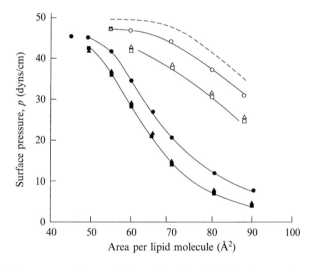

FIG. 6. Surface pressure (π)–area isotherms for phosphatidylserine monolayers formed at air–water (filled symbols) and oil–water interfaces (open symbols). The water phase was 0.1 M NaCl buffer, pH 7.4. The divalent cation concentrations were the same as those of the threshold concentration to induce vesicle fusion. The circles refer to the case without divalent cations, the squares refer to the case of the subphase solution containing 1 mM Ca^{2+}, and the triangles refer to the case containing 6 mM Mg^{2+}. From Ref. 26.

1–2 cm) of the inner wall of the glass trough should be treated with hydrophobizing chemicals (e.g., dimethyldichlorosilane) so that the glass surface of that portion is hydrophobic. First, an aqueous solution (e.g., 0.1 M NaCl–10 mM HEPES, pH 7.0) is placed in the glass trough up to the level of the hydrophobic glass surface. Then, a Wilhelmy plate is placed into the aqueous phase and the surface tension of the aqueous phase is measured. An aliquot of the lipid monolayer forming solution is placed onto the aqueous surface in small increments to spread a monolayer until a desired surface pressure (or area per molecule) is attained, and then the surface tension of the lipid monolayer is measured. After the measurement of the surface tension of the monolayer, hexadecane is placed onto the monolayer carefully and gradually (at least 1 cm depth of the oil phase), such that the Wilhelmy plate is immersed completely in the oil phase. Without a lipid monolayer, the interfacial tension of the oil–aqueous interface is approximately 50 dyn/cm. The pressure of the monolayer at the oil/water interface is defined as $\pi = \gamma_{o-w} - \gamma$, where γ_{o-w} and γ are the interfacial tension of the oil–water interface and that of the monolayer formed at the oil–water interface. Examples[24,26] of such experimental data are shown in Fig. 6, together with the results for the monolayers at the air–water surface.

Determination of Surface Hydrophobicity of Vesicle Membranes

The hydrophobicity, p, of the monolayer surface is defined as

$$p = \gamma/\gamma_{o-w} = (50 - \pi)/50 \tag{2}$$

where for $p = 1$, the monolayer surface corresponds to a hydrocarbon nature, and for $p = 0$, it corresponds to a water nature. From the $\pi–A$ relation for a given phospholipid, we can estimate the surface hydrophobicity of liposome membranes, knowing the area per lipid molecule for the liposome concerned. For example, for phosphatidylserine SUV liposomes, because the area per PS molecule of the PS membrane is ~80 Å^{2},[27] the surface hydrophobicity of the PS SUV liposomes is deduced to be approximately $p = 0.16$ ($50 - \pi = 8$ dyn/cm) from the corrected $\pi–A$ isotherm (a dashed line, which is corrected by considering the experimental values[28,29] of the film tension of lipid membranes) shown in Fig. 6. The absolute value of

[26] S. Ohki, in "Molecular Mechanisms of Membrane Fusion" (S. Ohki, D. Doyle, T. D. Flanagan, S. W. Hui, and E. Mayhew, eds.), p. 123. Plenum Press, New York, 1988.

[27] C. G. Brouillette, J. P. Segrest. T. C. Ng, and J. L. Jones, *Biochemistry* **21**, 4569 (1982).

[28] T. Tien, "Bilayer Lipid Membranes: Theory and Practice." Marcel Dekker, New York, 1974.

[29] E. Evans and D. Needham, in "Molecular Mechanisms of Membrane Fusion" (S. Ohki, D. Doyle, T. Flanagan, S. W. Hui, and E. Mayhew, eds.), p. 83. Plenum Press, New York, 1988.

surface hydrophobicity of the liposome membrane surface deduced by this method may have some variance because of various factors: (1) contamination of surface-active materials in the oil (hexadecane) used, (2) the method of forming the monolayer at the oil–water interface (a traditional method to obtain a π–A isotherm of the oil–water monolayer by compressing the monolayer is described elsewhere[30]; however, this method also involves variance in its value because of the impurity of the oil used and it will be more difficult to change the subphase solution condition properly by adding ions and others), and (3) the choice of the area per molecule for liposomes, because the area per molecule for lipid membranes varies, depending on the methods used to determine it.

In spite of these difficulties, however, the method presently described is reliable and useful to measure the change in surface hydrophobicity of the membrane with changes in environmental conditions, such as ionic concentrations in the aqueous phase. For example, the addition of small amounts of polyvalent cations results in a change in the interfacial tension of the monolayer from the reference point, which is interpreted as the change in surface hydrophobicity. Such studies have been performed with fusogenic ions. It has been shown that the addition of fusogenic ions in the aqueous phase alters the interfacial tension, increasing from the reference point (e.g., without fusogenic ions), depending on the concentrations and ionic species applied (Fig. 7).[25,26] It is also noted that as long as only the change (from the reference point) in hydrophobicity (or interfacial tension) due to the change in salt concentrations in the aqueous phase is concerned, both monolayer systems (monolayers formed at the air–aqueous phase and those formed at the oil–aqueous interface) give approximately the same results[26] (Fig. 6).

Contact Angle Method

Interfacial Tension Measurements by Contact Angle Method

The interfacial tension of a liposome membrane facing an aqueous phase can be obtained by measuring the contact angles of a set of liquids on liposome films and using a proposed theory.[23]

Theory

When an aqueous droplet is placed on a surface film, depending on its surface tension, the droplet contacts at an angle with respect to the film surface. The surface forces acting at the liquid contact point are in mechanical balance at an equilibrium state as illustrated in Fig. 8.

---

[30] J. H. Brooks and B. A. Pethica, *Trans. Faraday Soc.* **69**, 208 (1964).

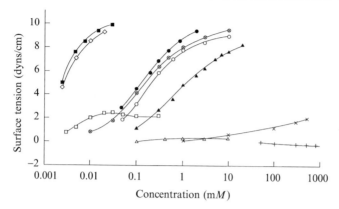

Fig. 7. The increase in interfacial tension of a phosphatidylserine monolayer (70 Å² per molecule) formed at the air–water [0.1 M NaCl, 1 mM histidine, 1 mM TES, 0.01 mM EDTA (0 mM EDTA for La³⁺ and Tb³⁺), pH 7.4] interface as a function of various metal cation and polyamine concentrations. ◇, La³⁺; ■, Tb³⁺; ●, Mn²⁺; ○, Ca²⁺; ▲, Mg²⁺; ⊗, H⁺; ×, Li⁺; +, Na⁺; □, spermine⁴⁺; △, spermidine³⁺. From Ref. 26.

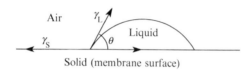

Fig. 8. A schematic diagram of a liquid drop placed on a solid (lipid membrane) surface.

The force balance at the contact point of liquid on the solid surface is expressed by Young's relationship[22]:

$$\gamma_S = \gamma_{S/L} + \gamma_L \cos\theta \tag{3}$$

where γ_S, γ_L, and $\gamma_{S/L}$ are the surface tensions of surface film (S) and liquid (L), and the interfacial tension of the surface film and the liquid, respectively, and θ is the contact angle.

Because the adhesion energy of two condensed phases, i and j, ΔG_{ij}, is given[31] by

$$\Delta G_{ij} = \gamma_{ij} - \gamma_i - \gamma_j \tag{4}$$

[31] J. Shaw, "Introduction to Colloid and Surface Chemistry," p. 130. Butterworths, London, 1980.

Eq. (3) can be rewritten as the Young–Dupré equation[31]:

$$(1 + \cos\theta)\gamma_L = -\Delta G_{SL} \tag{5}$$

According to the theory,[23] the surface tension, γ_i, of condensed phase material is produced by an apolar, Lifshitz–van der Waals (LW) component, and a polar Lewis acid–base (AB) component:

$$\gamma_i = \gamma_i^{LW} + \gamma_i^{AB} \tag{6}$$

where $\gamma_i^{AB} = 2\gamma_i^{\oplus}\gamma_i^{\ominus}$ and $\gamma_i^{\oplus}$ is the electron acceptor parameter and $\gamma_i^{\ominus}$ is the electron donor parameter of the polar component of the surface tension, γ_i^{AB}. These surface tension components and parameters define the interfacial tension between two condensed phase materials, i and j, as follows:

$$\gamma_{ij} = \left[\left(\gamma_i^{LW}\right)^{1/2} - \left(\gamma_j^{LW}\right)^{1/2} \right]^2$$
$$+ 2\left[\left(\gamma_i^{\oplus}\gamma_i^{\ominus}\right)^{1/2} + \left(\gamma_j^{\oplus}\gamma_j^{\ominus}\right)^{1/2} - \left(\gamma_i^{\oplus}\gamma_j^{\ominus}\right)^{1/2} - \left(\gamma_i^{\ominus}\gamma_j^{\oplus}\right)^{1/2} \right] \tag{7}$$

Here, γ_i^{LW}, $\gamma_i^{\oplus}$, and $\gamma_j^{\ominus}$ of a solid material, S, are determined by contact angle (θ) measurements, using the following version of the Young–Dupré equation:

$$(1 + \cos\theta)\gamma_L = 2\left[\left(\gamma_S^{LW} \cdot \gamma_L^{LW}\right)^{1/2} + \left(\gamma_S^{\oplus}\gamma_L^{\ominus}\right)^{1/2} + \left(\gamma_S^{\ominus}\gamma_L^{\oplus}\right)^{1/2} \right] \tag{8}$$

where subscripts S and L stand for "solid" and "liquid," respectively. The contact angle liquids most used for these measurements are listed in Table II. To determine the value for γ_S^{LW}, $\gamma_S^{\oplus}$, and $\gamma_S^{\ominus}$ on a flat, smooth surface of a solid material, it is necessary to use Eq. (8) three times: once to determine γ_S^{LW}, preferably by means of the contact angle with diiodomethane, and two other measurements, for example, with water and glycerol, in order to obtain the interfacial tension between a solid membrane film and water.

TABLE II
PROPERTIES OF CONTACT ANGLE LIQUIDS AT $20°$ C, IN MJ/M^2

Contact angle (liquid)	γ_L	γ_L^{LW}	γ_L^{AB}	$\gamma_L^{\oplus}$	$\gamma_L^{\ominus}$
Diiodomethane	50.8	50.8	0	0.01	0
Water	72.8	21.8	51.0	25.5	25.5
Glycerol	64.0	34.0	30.0	3.92	57.4
Formamide	58.0	39.0	19.0	2.28	39.6
Ethylene glycol	48.0	29.0	19.0	1.92	47.0

Data from van Oss, 1994.[23]

Methods and Procedures

Liposomes. Multilamellar phospholipid vesicles are prepared by hydrating dried phospholipids with 0.1 M NaCl buffer, pH 7.0, at 20 mg/ml and vortexed for 15 min at room temperature.

Preparations for Contact Angle Measurements. For contact angle (θ) measurements, layers of phospholipid vesicles are prepared by depositing a suspension of the multilamellar vesicles onto 0.2-μm-diameter pore size Hytrex silver membranes (Osmonics, Minnetonka, MN), by vaccum suction filtration. The number of vesicle membrane layers on the silver membranes can be about 250–500 layers. After filtration, the membranes are allowed to dry in air at 24° for about 40 min to reach the plateau period (of about 1 to 2 h), during which a constant value for the water contact angle is obtained ($\theta = 20°$).[32] All contact angle measurements should be done during the plateau period. Distilled water, including deionized and organic removal treatments, should be used, and pure salts should be used for preparation of various concentrations of salt solution.

Contact Angle Measurements. Contact angles can be measured with an NRL telescope/goniometer (Rame-Hart, Mountain Lakes, NJ). Liquid drops ($\sim$10 μl) are deposited on the vesicle layers above the silver membranes by means of 2-ml Teflon–glass syringes (Gilmont, Barrington, IL) with a 20-gauge needle. The contact angles (in degrees) can then be determined by visual observation.

Some Experimental Results

The contact angles measured by the previously-described method are shown in Fig. 9 for various phospholipid liposome membranes as a function of Ca^{2+} concentration in 0.1 M NaCl liquid droplets.[32] It is clearly demonstrated that PA membranes become more hydrophobic than PS membranes at a given concentration of Ca^{2+}, and PC membranes do not exhibit a change in hydrophobicity until a high concentration of Ca^{2+} has been achieved. Such contact angle measurements have been made on various lipid membranes, using various cations. It was demonstrated that fusogenic cations in the solution made acidic lipid membranes more hydrophobic[8,26] and the degree of hydrophobicity paralleled qualitatively the degree of fusion between membranes of the same composition. However, with these types of membranes, the interfacial tension at the membrane–water interface could not be determined because successful contact angle measurements with liquids other than water listed in Table III could not be made: the contact angles with these liquids were not stable. It is possible that

---

[32] M. Mirza, Y. Guo, K. Arnold, C. J. van Oss, and S. Ohki, *J. Disper. Sci. Tech.* **19,** 951 (1998).

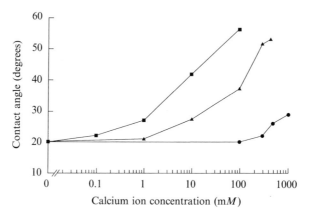

Fig. 9. Contact angles of $CaCl_2$ solutions placed on various phospholipid vesicle membranes as a function of Ca^{2+} concentration. ■, PA; ▲, PS; ●, PC. From Ref. 32.

similar experiments using oriented lipid monolayers formed on a hydrophobic solid (e.g., Teflon) surface may be successful in obtaining all information necessary to calculate the interfacial tension of a monolayer against the aqueous phase.

Some Comments and Possible Applications

Methods to determine the surface hydrophobicity of lipid membrane surfaces have been described, with measurements of the surface dielectric constant and the interfacial tension of the membranes. One of the situations in which these quantities become important is in the cases of strong adhesion and fusion of membranes. Normally, liposomes do not adhere strongly (at short distances) because of the high hydrophilicity of the membrane surfaces, creating structured water layers on the membrane surface as discussed in the Introduction. However, if these surfaces are altered by binding of polyvalent cations or dehydrated by PEG to a certain degree of hydrophobicity, liposomes can adhere strongly at short distances. Experimental results[6,15–17] and theoretical analyses[6] in this regard have been presented by several workers. Furthermore, if the adhesion of membranes becomes stronger, it causes deformation of the adhered membranes. Then, fusion of the membranes may occur through such deformed regions of the membranes, which are in a high-energy state and not stable.[33] The point at which the deformation induces a greater extent of membrane fusion, may

[33] S. Ohki and K. Arnold, *Colloids Surfaces B Biointerfaces* **18**, 83 (2000).

TABLE III

FUSION THRESHOLD CONCENTRATION OF CATIONS FOR VARIOUS LIPID VESICLES AND THEIR
ASSOCIATED MEMBRANE PROPERTIES[a]

Added ions in 0.1 M NaCl	Types of vesicles	Fusion threshold conc. of ions (mM)	Surface dielectric constant	Increased interfacial tension (dyns/cm)
	PS vesicles			
None	SUV	None	30	Reference value
	LUV	None	24[b,c]	Reference value
Na$^+$	SUV	None (up to 1 M)	30	0
	LUV	None (up to 1 M)	24[b,c]	0
K$^+$	SUV	None (up to 1 M)	30	0
H$^+$	SUV	2[d]	12	7
Ca^{2+}	SUV	1	12	7.5
	LUV	2.5[b,c,e]	14	7.5
Mg^{2+}	SUV	7	14	6.5
	LUV	None (up to 0.1 M)	-	-
Mn^{2+}	SUV	0.6	12	7.5
La^{3+}	SUV	0.004	12	7.5
Spermine^{4+}	SUV	None		A few dyns
	PA vesicles			
None	SUV	None	32	Reference value
Na$^+$	SUV	None (up to 1 M)	32	0
	LUV	None (up to 1 M)	24	0
Ca^{2+}	SUV	0.4[d]	14	7
	LUV	2.0[c]	13	7
Mg^{2+}	SUV	0.4[d]	14	7
	LUV	None (up to 0.1 M)	-	<6[d]
La^{3+}	SUV	0.004		
	PI vesicles[f]			
None	SUV	None	37	Reference value
Na$^+$	SUV	None	37	-
Ca^{2+}	SUV	10	12.5	-
	LUV	None	-	-
Mg^{2+}	SUV	90	-	-
Spermine^{4+}	SUV	None	-	-

[a] Data from Ohki and Arnold (1990) and Ohki (1988).
[b] Data from Köhler et al. (1997) and Binder et al. (2000).
[c] Data from Zschörnig et al. (unpublished data).
[d] Data from Ohki and Zschörnig (1993).
[e] Data from Wilschut et al. (1981).
[f] Data from Müller et al. (2003).

be called the fusion threshold condition. If such a condition is induced by
an elevated concentration of a fusogenic ion, the latter is termed the fusion
threshold concentration of the ion. Studies to determine fusion threshold
concentrations of various fusogenic ions for various liposomes have been

performed and such ion concentrations are listed in Table III, together with the values of dielectric constants and the increased interfacial tension corresponding to each threshold concentration. The surface tension increase corresponding to the fusion threshold for a given lipid membrane is nearly the same with a variety of ions.[8,24,33] Also, the surface dielectric constant corresponding to such a threshold point is approximately the same value.[8] It has been noticed that the magnitudes of the surface dielectric constant and the interfacial tension of the surface against the aqueous phase have a somewhat inverse relationship: the increase in one of the quantities relates to the decrease or reduction of the other. It was found that, except for some substances, the products of the two quantities are approximately constant (80–100 dyn/cm).[34] However, this relationship needs to be further studied to establish its validity. It should also be noted that the value of the constant should be different, depending on the method used to measure the dielectric constant of the membrane surface, because the surface dielectric constant varies with the surface region where it is measured, as discussed in the Introduction.

[34] S. Ohki, *Studia Biophys.* **127**, 89 (1988).

[17] Enzymatic Assays for Quality Control and Pharmacokinetics of Liposome Formulations: Comparison with Nonenzymatic Conventional Methodologies

By Hilary Shmeeda, Simcha Even-Chen, Reuma Honen, Rivka Cohen, Carmela Weintraub, and Yechezkel Barenholz

Introduction

Although extensive work has been performed to characterize liposomes, formal quality control (QC) of liposomal preparations has yet to be established and standardized (for reviews see Refs. 1–4). Quality control of liposomal dispersions is complicated by the supramolecular nature of the systems to be analyzed. The various components are held together by weak interactions to form lipid vesicles of the desired size and lamellarity. All chemical analyses of liposome lipid components require extraction of the lipid components, which must be complete, nonselective, and enable analysis and quantification of each of the components. In biological fluids, characterization of liposomal dispersions is further complicated because of the complexity of the milieu. Most assays monitoring stability are based on high-performance liquid chromatography (HPLC), however, quantification of lipids by HPLC suffers from major drawbacks,[1,2,4] being labor intensive, and requiring relatively expensive equipment (such as a differential refractometer or light-scattering detector) and standardized quantification.[1,2] Simple and reliable alternatives would be an important contribution to this area of study. Therefore, it is important to develop assay protocols that bypass the need for extraction, so that QC can be performed in any standard laboratory in the field of life sciences, pharmacy, or agriculture. These assays must be simple, reliable, not labor intensive, and with high throughput.

This chapter focuses on the use of commercial enzymatic kits for quantification of serum lipids and adaptation of these kits for liposome QC. These assays can be adapted conveniently in any laboratory for QC of liposomal formulations and of liposomal raw materials, and for follow-up of liposomes in biological fluids. These assays require only a simple spectrophotometer or, preferably, a high-throughput multiplate absorbance reader equipped with a monochromator or with filters of suitable wavelengths. Using the multiplate absorbance reader, the assays can be automated without sacrificing sensitivity, precision, and accuracy.

General Principles

The commercial kits relevant for liposome QC include kits for the determination of cholesterol and cholesterol esters; choline-containing phospholipids, including phosphatidylcholines, lysophosphatidylcholines, and sphingomyelins (which are major components of most current liposomal formulations); and nonesterified (free) fatty acids (NEFA), the major product of acyl ester lipid hydrolysis.[1,5] The latter two kits are used less

[1] Y. Barenholz and S. Amselem, *in* "Liposome Technology, 2nd Ed., Vol. 1, Liposome Preparation and Related Techniques" (G. Gregoriadis, ed.), pp. 527–616, CRC Press, Boca Raton, Florida, 1993.

[2] Y. Barenholz, Quality control of liposomes, *Special issue of Chem. Phys. Lipids* **64** (1993).

[3] R. R. C. New, "Liposomes: A Practical Approach", IRL Press, Oxford, 1990.

[4] S. W. Burgess, J. D. Moore, and W. A. Shaw, *in* "Handbook of Nonmedical Application of Liposomes", Vol. III (Y. Barenholz and D. D. Lasic, eds.), pp. 5–21. CRC Press, Boca Raton, Florida, 1996.

[5] D. Lichtenberg and Y. Barenholz, *in*, "Methods of Biochemical Analysis" (D. Glick, ed.), pp. 337–462. Wiley, New York, **33**, p. 337, 1988.

frequently. Although the commercial kits are convenient, homemade kits may be less expensive and can be assembled easily.[1,6] The commercial enzymatic kits are based on the stoichiometric generation of hydrogen peroxide by a specific oxidase and subsequent coupling of this to a chromogen, which, on oxidation via a peroxidase-catalyzed reaction, forms a chromophore in an amount proportional to that of the lipid analyte. Here we describe the assays that we have adapted for quantification on a multiplate absorbance reader. The spectra of the various chromophores formed in the three enzymatic assays described here are shown in Fig. 1.

The enzymatic assays in these kits are based on the use of a large excess of the specific enzymes so that the reaction is completed after a short incubation time. This requires complete exposure of the analyte to the enzyme. Therefore the kits we discuss include detergents that may be sufficient to solubilize plasma lipoproteins but in many cases are not sufficient for complete solubilization of liposomes. Raw materials such as phosphatidylcholine, sphingomyelin, and cholesterol should be solubilized in a suitable fresh detergent solution or dissolved in a solvent first. In addition, because these assays are based on the specific absorbance of the chromophore, the incubation mixture must be free of irrelevant absorbance or turbidity (at the wavelength of the chromophore) that cannot be corrected for by simple subtraction. In many (but not all) cases, turbidity can be eliminated by solubilization with a suitable detergent (see later). Samples should be measured for the possible contribution of extraneous turbidity by measuring the absorbance at a wavelength at which there is no contribution of the chromophore. Using a plate or multiplate absorbance reader under the conditions described in Table I, we recommend a wavelength of 690 nm.

General Advantages of Enzymatic Assays

- No need for extraction or complex handling
- High throughput
- Small samples
- High sensitivity

General Limitations of the Use of the Enzymatic Assays

- Samples must be diluted to suit the concentration range of the assays, as described in Table I.
- The use of multiplate absorbance microwells requires accurate dilution of samples.

---

[6] H. U. Bergmeyer, ed., "Methods in Enzymatic Analysis, Vol. 8, Lipids," VCH, Weinheim, Germany, 1985.

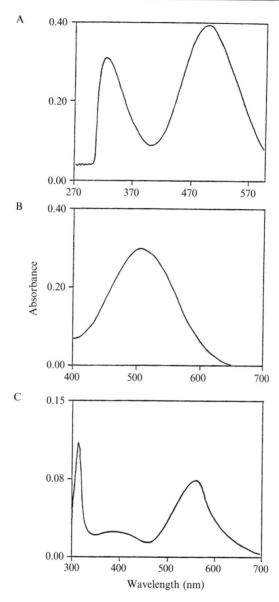

Fig. 1. Spectra of the chromophores produced in the three enzymatic assay kits: (A) nonesterified cholesterol, "Nonesterified Cholesterol" kit, Boehringer Mannheim; (B) choline phospholipid, "Phospholipids B" kit, Wako Chemicals; (C) nonesterified fatty acids, "NEFA C" kit, Wako Chemicals.

TABLE I
PREPARATION OF CHOLESTEROL STANDARD CURVE

Stock aliquot (μl)	Isopropanol (μl)	Working standard concentration (nmol/μl)	Nanomoles per assay
0	1000	0	0
100	900	1	10
200	800	2	20
300	700	3	30
400	600	4	40
600	400	6	60
800	200	8	80

- Any agent or side reaction that will affect the level of hydrogen peroxide will also interfere with the assay results. This can be tested by measuring the sample directly for the presence of hydrogen peroxide, based on a simple assay described by Pick and Keisari.[7] This assay measures the horseradish peroxidase–mediated oxidation of phenol red by hydrogen peroxide, which generates a compound that can be detected at a wavelength of 610 nm. Alternatively, it is possible to use catalase to induce hydrogen peroxide degradation, a process that can be monitored spectrophotometrically at 240 nm.[8]

Some reactive species in the analyte mixture, for example, the nitroxide stable radicals,[9] have the ability to interact directly with the chromogen 4-aminoantipyrine. Appropriate controls should be run when such reactions are anticipated (see Determination of Nonesterified Fatty Acids).

- The use of commercial kits requires strict adherence to the manufacturer's instructions, including storage conditions of the reagents and samples, sample preparation, and calibration.
- Cloudy (turbid) samples will interfere with absorbance. *Note:* Extracted samples may produce a cloudy dispersion after resuspension. In the NEFA assay this can be overcome by using the detergent

[7] E. Pick and Y. Keisari, *J. Immunol. Meth.* **38,** 161 (1980).
[8] B. Chance and C. A. Maehly, *in* "Methods of Enzymology, Vol. II" (S. P. Colowick and N. O. Kaplan, eds.), p. 764. Academic Press, New York, 1955.
[9] J. B. Mitchell, M. C. Krishna, A. Samuni, A. Russo, and S. M. Hahn, *in* "Reactive Oxygen Species in Biological Systems" (D. L. Gilbert and C. A. Colton, eds.), pp. 293–313. Kluwer Academic Press/Plenum Publishers, New York, p. 293 (1999).

Triton X-100, which will not interfere with the assay results. Usually complete solubilization is obtained at a detergent concentration equal to its critical micellar concentration (CMC) plus twice the concentration of total lipid.[1]

• When serum is analyzed, hemolytic blood samples may interfere with the assay (see below).

Determination of Nonesterified Cholesterol

Background

The following protocol is a modification of the Free Cholesterol kit (Wako Chemicals, Neuss, Germany) protocol, adapted for evaluation of many samples in a multiplate absorbance reader. Commercial kits are available from a variety of manufacturers; similar results are obtained with kits from Boehringer Mannheim (Mannheim, Germany) and Sigma (St. Louis, MO). The enzymatic assay kit for determination of nonesterified (free) cholesterol is based on the following reaction sequence:

$$\text{Cholesterol} + O_2 \xrightarrow{\text{cholesterol oxidase}} \text{cholesten-4-one} + H_2O_2$$

$$2H_2O_2 + \text{4-aminophenazone} + \text{phenol} \xrightarrow{\text{peroxidase}}$$
$$\text{4-}(p\text{-benzoquinone-monoimino)-phenazone} + 4H_2O$$

The quinoneimine reaction product is measured by its absorbance at 500 nm,[4,5] using a spectrophotometer or a plate reader with a filter that allows quantifying light at $\geq$500 nm.

A separate kit is available for total cholesterol. This includes the following reaction that converts esterified cholesterol to nonesterified cholesterol:

$$\text{Cholesterol ester} + H_2O \xrightarrow{\text{cholesterol esterase}} \text{cholesterol} + \text{RCOOH}$$

Esterified cholesterol is calculated as the difference between the amount of total and free cholesterol.

Assay Protocol

Cholesterol Reagent

Buffer 1: 0.4 M potassium phosphate (pH 7.7), 20 mM phenol, 1.85 M methanol

Buffer 2: 0.4 M potassium phosphate (pH 7.7), 2 mM 4-aminophenazone, 1.8 M methanol, 0.4% hydroxypolyethoxydodecane

Solution 3: Cholesterol oxidase (12 U/ml), peroxidase (8 U/ml)

To 1 volume of buffer 2, a 1/10 volume of solution 3 is pipetted and 1 volume of buffer 1 is added (e.g., 0.5 ml of solution 3 is pipetted into 5 ml of buffer 2, and then 5 ml of buffer 1 is added). This solution is stable for 2 weeks at 4°.

Cholesterol Standard Stock Preparation. The standard stock solution is prepared at a concentration of 10 nmol/μl. Cholesterol (99.0%; Sigma) (38.66 mg) is weighed and dissolved in 10 ml of warm HPLC-grade isopropanol. This is stored in aliquots of 2 ml at −20°.

Working Standard Solution of Cholesterol (10–80 nmol/220-μl assay). The cholesterol stock standard (10 nmol/μl, stored at −20°) is warmed to room temperature. To make working standard solutions, aliquots of thawed stock standard are diluted with isopropanol as shown in Table I.

Assay Procedure

1. Ten microliters of each standard concentration is pipetted into the wells of a 96-well microplate. Saline is added to a final volume of 20 μl. *Note*: 10 μl of isopropanol does not affect the assay results.

2. Two hundred microliters of cholesterol reagent is pipetted into the microwells.

3. An appropriate amount of sample is pipetted to a final volume of 20 μl, for a final assay volume of 220 μl.

4. The microplate is vortexed gently or placed on a microplate shaker.

5. The microplate is incubated at 20–25° for 10 min or at 37° for 5 min.

6. The absorbance is read at 500 nm in a plate reader (for spectra, see Fig. 1). The equation describing the correlation for the standard curve is as follows: $y = 0.0183x - 0.0035$; $r^2 = 0.9989$, where y is absorbance at 500 nm, x is the amount of cholesterol (nanomoles per assay) in the sample, and r^2 is the correlation coefficient (Fig. 2B).

Comparison of Enzymatic and HPLC Cholesterol Determinations

The above-described method of determination has been compared with that of HPLC based on the protocol of Smith et al.[10] Figure 2A and B presents the standard curves for these two methodologies. The commercial kits (such as those from Boehringer Mannheim, Sigma, and Wako chemicals) are able to detect accurately small amounts of nonesterified cholesterol, as low as 1.5 nmol. The correlation coefficient is similar for both methods. The enzymatic assay is more sensitive than HPLC

[10] L. L. Smith, V. B. Smart, and G. A. Ansari, *Mutat. Res.* **68**, 23 (1979).

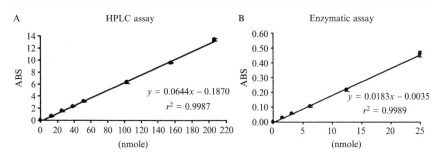

Fig. 2. Side-by-side comparison of free (nonesterified) cholesterol calibration curves obtained by HPLC (A) and by the enzymatic (cholesterol oxidase) assay (B).

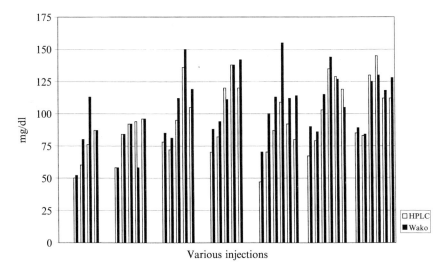

Fig. 3. Comparison of nonesterified cholesterol levels measured by the enzymatic Wako kit (■) or HPLC (□) in human plasma after intravenous infusion of egg PC SUV.

cholesterol determination. The presence of at least 100-fold phosphatidylcholine does not interfere with the enzymatic determination.

The two methods have been compared for quantitation of nonesterified cholesterol in human serum samples treated with a formulation of egg phosphatidylcholine small (∼100 nm) unilamellar vesicles (SUV), which causes a transient rise in serum cholesterol levels. The results demonstrate a good correlation (Fig. 3).

Quantification of Cholesterol by HPLC Normal-Phase Method

Cholesterol from the liposome must first be extracted for example, with Dole reagent (also see Barenholz and Amselem[1]) as follows: an aliquot of up to 0.25 ml of aqueous sample (completed this volume) containing liposomes is placed in an Eppendorf polypropylene test tube. Dole reagent, 0.5 ml (composed of 400 ml isopropanol: 100 ml heptane, both HPLC grade, and 10 ml of analytical grade or better 1 N H_2SO_4) is added and the test tube is vortexed for 10 s. Heptane, 0.25 ml is added and the mixture is vortexed for 30 s and centrifuged in an Eppendorf centrifuge for 2 min. Aliquots of the Dole heptane-rich upper phase are removed for the HPLC analyte. To prevent heptane evaporation, the test tubes are kept sealed and cold. The Dole procedure removes all polar lipids and nonlipids. The HPLC method is based on Smith *et al.*[10] with small modifications. An aliquot of cholesterol sample in heptane or of the heptane-rich Dole upper phase is separated in the HPLC system using a normal-phase, 3-μm particle size silica column, 100 × 4.6 mm (Alltech Associates, Deerfield, IL), in a mobile phase composed of hexane–isopropanol (94:6, v/v) at a flow rate of 1 ml/min, with spectrophotometric detection at 212 nm. Under such conditions, the cholesterol retention time is ∼3.0 min. Calibration curves are prepared under conditions identical to those for sample preparation. The sensitivity limit of this assay is 0.5 nmol of cholesterol (0.2 μg) per peak.

Determination of Choline Phospholipid Content

Background

A simple, rapid, and accurate kit (Phospholipids B kit) has been developed by Wako Chemicals (Neuss, Germany) for the determination of serum choline phospholipids. The method employs the enzyme phospholipase D, which hydrolyzes the ester bond between the phosphate and the choline of lecithin, lysolecithin, and sphingomyelin phospholipids. Their water-soluble derivatives, glycerophosphorylcholine and phosphorylcholine, are not hydrolyzed by phospholipase D. The free choline is then oxidized by choline oxidase to betaine, with quantitative release of hydrogen peroxide. Hydrogen peroxide, catalyzed by peroxidase, then couples 4-aminoantipyrine and phenol, generating a red product with maximum absorption at a wavelength of 505 nm. The reaction sequence is as follows:

$$\text{Phospholipids} \left\{ \begin{array}{l} \text{lecithin} \\ \text{sphingomyelin} \\ \text{lysolecithin} \end{array} \right\} + \text{H}_2\text{O} \xrightarrow{\text{phospholipase D}}$$

$$\text{choline} + \left\{ \begin{array}{l} \text{phosphatidic acid} \\ N\text{-acylsphingosyl phosphate} \\ \text{lysophosphatidic acid} \end{array} \right.$$

$$\text{Choline} + 2\text{O}_2 \xrightarrow{\text{choline oxidase}} 2\text{H}_2\text{O}_2 + \text{betaine 4-aminoantipyrine}$$
$$+ \text{phenol} + 2\text{H}_2\text{O}_2 \xrightarrow{\text{peroxidase}} \text{quinone pigment (red)} + 4\text{H}_2\text{O}$$

This method simplifies greatly the conventional methods, which generally involve three steps: (1) extraction, usually with solvents or by precipitation, (2) conversion of organic phosphorus to inorganic phosphorus, which is usually accomplished by wet acid digestion with perchloric and/or sulfuric acid and necessitating heating at high temperature, and (3) quantification of the inorganic phosphorus released, most commonly by color development (see later and Ref. 1).

Assay Protocol

Reagents: Phospholipids B Kit

Buffer solution: 0.05 M Tris buffer solution, pH 8.0, containing 0.05% phenol

Color reagent: Phospholipase D (0.4 U/ml), choline oxidase (2 U/ml), peroxidase (5 U/ml), and 0.015% 4-aminoantipyrine

Standard solution: Choline chloride (54 mg/dl, which corresponds to a phospholipid content of 3.87 mM or 300 mg/dl) and 0.1% phenol

Preparation of Standard Curve (See Table II)

Reagents, samples, and standards should be brought to room temperature before use.

Choline standard preparation: The kit uses choline for the standard curve. The standards ($\sim$2.0 to $\sim$26.0 nmol per assay) should be prepared in 0.05 M Tris buffer, pH 8.0, and diluted up to 1:10 for use in microwell plates, in a final volume of 20 μl (final volume can be adjusted with highly purified water), (resistance, 18.2 MΩ), obtained using purification systems such as the WaterPro PS HPLC/Ultrafilter hybrid system (Labconco, Kansas City, MO) or equivalent instrumentation that provides low levels of total organic carbon and inorganic ions in pyrogen-free sterile water. A liposomal dispersion of phosphatidylcholine that is quantified by a

TABLE II
CHOLINE PHOSPHOLIPID CALIBRATION CURVE

Standard diluted 1:10 (μl)	Water (μl)	Nanomoles per well
0	20	0
5	15	1.94
10	10	3.88
20	0	7.74
40[a]	—	14.54
Undiluted standard		
5	15	19.35
7.5	12.5	25.80

[a] Corrected for final volume, adjusted from 320 to 340 μl.

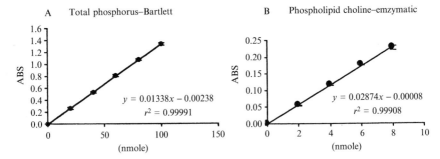

FIG. 4. Comparison of phosphatidylcholine in small unilamellar vesicles by (A) total phosphorus (modified Bartlett) method and by (B) enzymatic choline phospholipid determination (Wako).

modified Bartlett organic phosphorus determination[1] has been used by us successfully as an alternative standard.

The equation for the standard curve is as follows: $y = 0.02874x + 0.00008$; $r^2 = 0.99908$, where y is the absorbance at 505 nm, x is the amount of phospholipid (nanomoles per assay), and r^2 is the correlation coefficient (Fig. 4B). Include a standard/sample blank of either buffer or saline.

Assay Procedure.

1. Three hundred microliters of color reagent is added to each well to obtain a final volume per well of 320 μl.

2. The standard is added in a final volume of 20 μl. (For liposome, dispersions of 10% phospholipid, dilute 1:200 in saline.)

3. The plate is vortexed gently.

4. Samples are incubated at 37° until completion of the reaction (usually 10–30 min).

5. The absorbance of the colored product is read in a spectrophotometer at 505 nm, or in a plate reader equipped with a 500-nm filter. The colored product is stable for 2 h. Results are linear up to 26.0 nmol per assay and correlate with values quantitated by the modified Bartlett method[1] to within 5%.

Comparison of Enzymatic and Chemical Determinations

Standard Curves. The quantification of liposomal raw materials and dispersions on the basis of their phosphatidylcholine concentrations by the enzymatic Phospholipids B kit has been compared with that obtained by the modified Bartlett organic phosphorus method,[1] which determines organic phosphorus extracted in the Bligh and Dyer or the Folch chloroform-rich lower phase (included below). An excellent correlation has been observed between the two methods. The standard curves generated by each method are displayed in Fig. 4A and 4B. [Only highly pure water.]

Phospholipid Determination by a Modified Bartlett Procedure

The procedure is based on determination of the level of phosphorus, the common denominator for all phospholipids. To be specific for phospholipids it is necessary to extract the phospholipids from all other compounds that contain inorganic or organic phosphorus. Any procedure in which phospholipid recovery is complete (>95%) and reproducible can be used (for more details see Barenholz and Amselem[1]). We routinely use either the Folch or the Bligh and Dyer extraction procedure. Only high-purity water (see above) should be used.

Reagents

Reagent 1: 70–72% HClO$_4$ (phosphorus free; Merck)

Reagent 2: 5% ammonium molybdate solution in water (low-phosphorus ammonium molybdate; BDH, Poole, UK)

Reagent 3: Fiske and Subbarow reducer (Sigma), 0.5 g/3 ml in high-purity water

TABLE III
TOTAL PHOSPHOLIPID DETERMINATION AT THREE SENSITIVITY RANGES

Addition	Sensitivity range (nmol)		
	20–500	8–150	4–70
$HClO_4$[a]	1.0 ml	0.4 ml	0.2 ml
H_2O	3.3 ml	1.2 ml	0.6 ml
Ammonium molybdate[b]	0.6 ml	0.2 ml	0.1 ml
Reducer[c]	150 μl	50 μl	30 μl
Final volume	4.75 ml	1.73 ml	0.87 ml

[a] Reagent 1; see step 3 in Phospholipid Determination by Modified Bartlett Procedure.
[b] Reagent 2; see step 5 in Phospholipid Determination by Modified Bartlett Procedure.
[c] Reagent 3; see step 5 in Phospholipid Determination by Modified Bartlett Procedure.

Phosphate standard solution; potassium phosphate (Sigma), 0.65 mmol/l

Assay Procedure. Use disposable borosilicate glass 16 × 150 mm test tubes or, alternatively, glass test tubes washed with sulfochromic acid composed of 9 volumes of 3% potassium bichromate and 1 volume of concentrated H_2SO_4. The glassware cleanliness/purity is crucial to obtain a low background.

1. Standards/samples in aqueous phase should be adjusted to the final sample volume with highly pure H_2O based on the sensitivity range selected (see Table III). When the sample is dissolved in an organic solvent phase, the solvent should first be evaporated to dryness. **Note: This step is very important to eliminate the risk of an explosion.**

2. A few acid-washed silicon carbide boiling stones (Thomas Scientific, Swedesboro, NJ) are added to each sample to ensure safe boiling and to prevent loss.

3. Add $HClO_4$ (reagent 1) to fit the selected sensitivity range (Table III).

4. Heat to the boiling point (180–200°) for 30 min, and then cool to room temperature.

5. Water, then reagent 2, and then reagent 3 are added to fit the desired sensitivity range (Table III) and mixed well after each addition. Immediate mixing is important to obtain low and reproducible blanks because ammonium molybdate may contain a small amount of phosphorus.

6. The test tubes are heated at 100° for 7 min, and cooled to room temperature.

The samples are read in a spectrophotometer at 830 nm for high sensitivity or at 660 nm for low sensitivity, against the reagent blank. Each determination should be performed in triplicate and should include at least a partial calibration curve and reagent blanks. It is recommended that the user compare the reagent blanks and calibration curve between various experiments in order to detect deviations; high-purity water should be used as the analyte in the reagent blanks.

The majority of phospholipid classes used for the preparation of liposomes contain exactly 1 mole of phosphorus per mole of phospholipid; therefore, the phospholipid concentration can be derived directly from a quantification of the lipid phosphorus content of the sample. However, an exception to this rule is cardiolipin, which contains 2 moles of phosphorus per mole of phospholipid. Therefore, if the sample contains this lipid an appropriate correction must be made in the calculation. This modified Bartlett method covers a broad range of sensitivity by a combination of spectrophotometric determinations at 660 and 830 nm and volumes of 1 to 5 ml. The sensitivity range is 2–1000 nmol.

From our experience with the modified Bartlett assay it is clear that the assay is not trivial and requires training. Problems are due to extraction procedures, errors due to pipetting organic solvents, incomplete digestion, contamination of the reagents, improper mixing, and contamination of the glassware by phosphorus (even common dust contains phosphorus). Reproducible results can be obtained with sufficient care. A comparison between the two assays applied to egg phosphatidylcholine small unilamellar vesicles is shown in Fig. 4A and 4B.

Pharmacokinetics of Liposomal Phospholipids: Comparison of Enzymatic Assay with Organic Phosphorus Determination. The pharmacokinetics of intravenously administered liposomes containing sphingomyelin or phosphatidylcholine can be monitored by either the enzymatic assay or the modified Bartlett procedure, either with or without separation of the liposomes from plasma lipoproteins. We have compared the use of these two methods for determining the pharmacokinetics of egg phosphatidylcholine small unilamellar vesicles in human plasma (Fig. 5). Although there is general agreement, there are significant discrepancies between absolute levels detected by the two methodologies. The Bartlett results are generally higher, as expected, because it measures all the types of phospholipids in serum, whereas the enzymatic assay determines only choline phospholipids. However, the percent change from the baseline values due to phosphatidylcholine liposomes infused correlates reasonably well (with maximal differences of $\pm 28\%$). The enzymatic kit is by far easier and much less labor intensive than the modified Bartlett method.

Specific Limitations of Enzymatic Assay

1. The enzymatic assay determines only the three choline phospholipids sphingomyelin, phosphatidylcholine, and lysophosphatidylcholine,

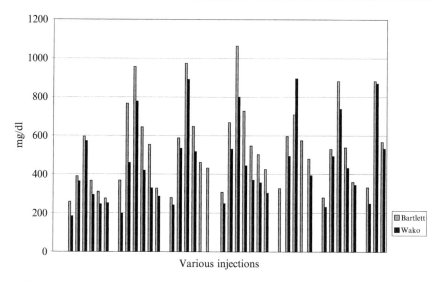

Fɪɢ. 5. Comparison of total phosphorus determined by the modified Bartlett method (▨) and total choline phospholipid by the enzymatic (Wako) kit (▪) in human plasma after intravenous infusion of egg PC SUV.

whereas the modified Bartlett procedure determines all lipids containing phosphorus.

2. Extracted samples sometimes present a turbidity problem that needs to be overcome.

3. The presence of antioxidants in serum or other compounds that may affect H_2O_2 may interfere with the enzymatic assay (see General Limitations).

Determination of Nonesterified Fatty Acids

Background

Liposomal instability commonly results from phospholipiol acyl ester hydrolysis, producing nonesterified fatty acids (NEFA). We have adapted the Wako NEFA C kit for nonesterified fatty acids in serum, to monitor liposome chemical stability.[10–13] This kit is based on the acylation of

[11] D. Simberg, D. R. Hirsch-Lerner, R. Nissim, and Y. Barenholz, *J. Liposome Res.* **10,** 1 (2000).

[12] I. Babai, S. Samira, Y. Barenholz, Z. Zakai-Rones, and E. Kedar, *Vaccine* **17,** 1223 (1999).

[13] A. Samuni, A. Lipman, and Y. Barenholz, *Chem. Phys. Lipids* **105,** 121 (2000).

coenzyme A by long-chain fatty acids, catalyzed by the enzyme acyl-CoA synthetase (ACS). The acyl-CoA is then oxidized quantitatively by acyl-coenzyme A oxidase (ACOD), generating 2,3-*trans*-enoyl-CoA and hydrogen peroxide. The quantitative generation of hydrogen peroxide is then utilized in the oxidative condensation of 3-methyl-N-ethyl-N-(β-hydroxyethyl)-aniline (MEHA) with 4-aminoantipyrine in the presence of peroxidase (POD), generating a purple adduct, proportional in quantity to that of NEFA. The adduct is quantified spectrophotometrically at 550 nm.

The reaction sequence is as follows:
First incubation:

$$RCOOH(NEFA) + ATP + CoA \xrightarrow{ACS} acyl\text{-}CoA + AMP + PP_i$$

Second incubation:

$$Acyl\text{-}CoA + O_2 \xrightarrow{ACOD} 2,3\text{-}trans\text{-}enoyl\text{-}CoA + H_2O_2$$

$$2H_2O_2 + MEHA + 4\text{-}aminoantipyrine \xrightarrow{POD} color\ adduct + 4H_2O$$

Advantages and Limitations

This is a unique kit because it enables determination of the level of free long-chain fatty acids despite the presence of high levels of other lipids, without the need for separation. Therefore, this method can be used to monitor hydrolysis of acyl esters from ester lipids.

Because of the high sensitivity of this assay, antioxidants may interfere with the reaction. For example, ascorbic acid present in serum may have significant antioxidant activity that may interfere with the formation of hydrogen peroxide. Ascorbate oxidase (AOD) is therefore included in reagent A of this kit to neutralize the effects of ascorbic acid before the oxidation of the acyl-CoA occurs. We have tested the effects of other antioxidants commonly encountered in liposomal formulations. α-Tocopherol at a level of 0.2 mol% of lipids does not interfere with the assay. Nitroxides, which are stable radicals that act as self-replenishing antioxidants,[9] interfere with the reaction at concentrations above 1 mM.

We have used this assay routinely and successfully for QC and to monitor the chemical stability of liposomal formulations during processing and storage.[1–5,11,14,15] Our results are in excellent agreement with other stability

[14] N. J. Zuidam and Y. Barenholz, *Biochim. Biophys. Acta* **1329,** 211 (1997).
[15] K. Shinoda, T. Nakagawa, B. Tamamushi, and T. Isemura, "Colloidal Surfactants," p. 8. Academic Press, New York, 1963.

indicators [HPLC and thin-layer chromatography (TLC)[11–14]] with the advantage that NEFA determination is much simpler and more reproducible (R. Cohen, H. Shmeeda, and Y. Barenholz, unpublished data, 2003). Once liposomes are solubilized, the assay is not affected by the type or lipid composition of the liposomes. SUV, large unilamellar vesicles (LUV), multilamellar vesicles (MLV), dehydration–rehydration vesicles (DRV), and other types of liposomes made of a broad spectrum of lipids, including saturated and unsaturated, zwitterionic, anionic, and cationic lipids, in the presence and absence of cholesterol, have been assayed successfully.[11–14] The presence of PEGylated lipids in the liposomes slows the rate of the first reaction; therefore, incubation must be prolonged from 10 to 20 min (see Assay Procedure, step 3). The NEFA levels of liposomal formulations containing drugs that do not interfere with the absorbance of the product spectrally (such as amphotericin B) have also been successfully determined. The presence of drugs that have absorbance overlapping with the assay, such as doxorubicin, requires the development of procedures to remove these drugs before the enzymatic NEFA assay is performed.

Assay Protocol

   Reagents
     Color reagent A: Provided in lyophilized form and containing

Acyl-coenzyme A synthetase (ACS)	3 U/vial
Ascorbate oxidase (AOD)	30 U/vial
Coenzyme A (CoA)	7 mg/vial
Adenosine triphosphate (ATP)	30 mg/vial
4-Aminoantipyrine	3 mg/vial

     Diluent for color reagent A: 65 ml of an aqueous solution containing

Phosphate buffer, pH 6.9	0.05 M
Magnesium chloride	3 M
Surfactant and stabilizers (unspecified)	

     Color reagent B: Provided in lyophilized form:

Acyl-coenzyme A oxidase (ACOD)	132 U/vial
Peroxidase (POD)	150 U/vial

     Diluent for color reagent B: 130 ml of an aqueous solution containing the following:
     (3-Methyl-*N*-ethyl-*N*-(β-hydroxyethyl)-aniline (MEHA)
                                        1.2 mmol/L surfactant
    NEFA standard solution: 10 ml of an aqueous solution of nonesterified fatty acid of known concentration:
     Oleic acid (1.0 mM; 1.0 m Eq/L)

Surfactant and stabilizers

Triton X-100 (20%): It is important to prepare a fresh solution of this in doubly distilled water before use

Preparation of Reagent Solutions

Color reagent A solution: 10 ml of diluent for color reagent A is added to 1 vial of dry color reagent A and mixed gently by inversion until completely dissolved

Color reagent B solution: 20 ml of diluent for color reagent B is added to one vial of color reagent B and mixed gently by inversion until completely dissolved

Liposome Solubilization. Liposome samples containing 1% lipid (by weight) are dissolved in freshly prepared 20% Triton X-100 [a detergent concentration of at least the detergent CMC + (2 × total lipid concentration)[1] is needed]. For turbid samples 20% Triton X-100 is added and the samples are heated at a temperature above the "cloud point," and then cooled at once to 4°. The cloud point is defined as the temperature at which phase separation of nonionic detergent occurs due to temperature elevation (which leads to dehydration and an increase in the micelle aggregation number).[15]

Assay Procedure

1. The kit standard is diluted 1:10 with doubly distilled water: 15, 30, 45, and 60 μl are pipetted for a linear range of 1.5–6 nmol of fatty acid (Table IV). All samples are adjusted to a final volume of 60 μl. (*Note:* Use of detergent for solubilization of samples requires addition of the detergent to the standards for calibration.)

The absorbance of the reagent blank containing Triton X-100 should be between 0.060 and 0.150 OD units. The equation describing the correlation of the standard curve is as follows: $y = 0.0382x + 0.0032$; $r^2 = 0.995$, where y is the absorbance at 540 nm, x is the concentration of fatty acid in nanomoles per 260 μl, and r^2 is the correlation coefficient. The presence of 20% Triton X-100 does not significantly affect this correlation ($<5.0\%$). A standard/sample blank of doubly distilled water is included in the assay.

2. Sixty microliters of lipid sample is pipetted into 96-well microplates. To turbid samples 20 μl of freshly prepared 20% Triton X-100 is added. If turbidity is not resolved by the addition of Triton X-100, the sample is warmed to the cloud point and then cooled in order to solubilize the liposomes. For nonturbid samples, 20 μl of doubly distilled water is added for a final sample volume of 80 μl.

TABLE IV
NEFA CALIBRATION CURVE

Standard diluted 1.10 (μl)	Water (μl)	Nanomoles NEFA per well
0	60	0
15	45	1.5
30	30	3.0
45	15	4.5
60	0	6.0

3. Sixty microliters of reagent A is pipetted, mixed well, and incubated for 10 min at 37°. For liposomes containing PEGylated lipids, the incubation time must be extended to 20 min (see Advantages and Limitations).

4. One hundred and twenty microliters of reagent B is added, mixed well, and incubated for 10 min at 37° for a final volume of 260 μl.

5. The absorbance is read at 540 nm for NEFA and at 690 nm to measure turbidity interference.

Identical slopes and absolute values were obtained for NEFA alone or in the presence of a large excess (1.5 μmol) of the following phospholipids: (1) dimyristoyl phosphatidylcholine (DMPC), (2) DMPC–dimyristoylphosphatidylglycerol (DMPG) (9:1), and (3) N-(1-(2,3-dioleoyloxy)-propyl)-N,N,N-trimethylammonium chloride (DOTAP).

Specific Limitations

1. Samples that are icteric, hemolyzed, or lipemic may result in inaccurate results, and a specimen blank must be included.

2. Samples suspected or known to have compounds that may directly interact with the chromogen should be incubated directly with reagent B as a control.

3. Heparin activates plasma lipoprotein lipase, which then releases NEFA of triacylglycerides and phospholipids; therefore plasma samples containing heparin are not suitable for this assay.

Conclusion

We have detailed a number of methods for the adaptation of commercially available enzyme-based kits for lipid analysis that are suitable for quality control and pharmacokinetics of liposomal formulations. A summary of these appears in Table V. We hope that this will facilitate the

TABLE V
ENZYMATIC ASSAYS

Kit for determination of:	Specific enzyme I	Specific enzyme II	Wavelength of filter[a] (nm)	Wavelength of maximum absorbance (nm)	Sensitivity range (nmols)	Linear regression[b] $y = ax + b$		
						a	b	r^2
Nonesterified cholesterol	Cholesterol oxidase	None	500	505	1.5–80	0.018	0.004	0.999
Choline phospholipids	Phospholipase D	Choline oxidase	500	505	2.0–30.0	0.029	0.000	0.999
Nonesterified fatty acids	Acyl-CoA synthetase	Acyl-CoA oxidase	540	550	1.5–6.0	0.0382	0.0032	0.995

[a] For use of absorbance plate reader equipped with filters.
[b] y is the absorbance (OD), a is the slope, x is nanomoles of analyte per assay, b is the y intercept, and r^2 is the correlation coefficient.

development of common assay protocols and allow standardization of lipid and liposome quality control assays.

Acknowledgments

The student fellowship support of Lipoid GmbH (Ludwigshafen, Germany) and of the US–Israel Binational Science Foundation for development of choline phospholipid determination, Sequus Pharmaceuticals (now Alza, Menlo Park, CA) support for the development of the nonesterified fatty acid assay, and the Israel Science Foundation grant for development of the nonesterified cholesterol assay are gratefully acknowledged. We appreciate the help of Mr. Sigmund Geller in editing the manuscript and the help of Mrs. Beryl Levene in typing it.

Author Index

Subject Index

A

AFM, *see* Atomic force microscopy
Atomic force microscopy
  advantages in liposome imaging, 201
  buffers for liposome imaging, 205
  contact mode, 207–208
  image presentation and interpretation,
    208–209
  instrumentation, 199–200
  liposome preparation
    imaging in air sample preparation,
      203–204
    imaging in liquid sample preparation,
      205–207
    lipid composition, 202–203
    polymerization of liposomes, 203
    size effects, 202–203
  principles, 199
  real-time imaging of glycolipid-containing
    liposomes, 209–213
  substrates for liposome immobilization
    selection, 201–202
    silicon cleaning, 202
  tapping mode, 208

C

Capillary viscometry, *see* Viscometry
Cholesterol, enzymatic determination of
    nonesterified cholesterol for liposome
    quality control
  advantages and limitations, 274, 276
  commercial kits, 273–274
  comparison with high-performance
    liquid chromatography assay,
      278–280
  controls, 276–277
  materials, 277–278
  microtiter plate preparation and reading,
    277–278
  principles, 277, 290

Choline phospholipids, enzymatic
    determination for liposome quality
    control
  advantages and limitations, 274, 276,
    285–286
  commercial kits, 273–274
  comparison with modified Bartlett
    phosphorous determination,
      283–285
  controls, 276–277
  materials, 281
  principles, 280–281, 290
  sensitivity, 290
  standard curve preparation, 281–283
CMC, *see* Critical micellar concentration
Critical micellar concentration, detergents,
  48–50

D

Dansylhexadecylamine, fluorescence
    anisotropy studies of packing and
    polarity change, 229–230
Dehydration–rehydration vesicles, *see*
    Plasmid DNA liposome vaccine
Detergent removal, liposome
    preparation
  advantages, 47
  bead adsorption, 68
  detergents
    critical micellar concentrations, 48–50
    selection factors, 48
  dialysis
    flat membranes, 65–66
    hollow fibers, 66–67
  dilution, 64
  filtration, 67–68
  gel chromatography, 64–65
  limitations, 47
  mixed micelles
    aggregation behavior, 50, 52
    coexistence with vesicles, 54

AHL AND PERKINS, CHAPTER 7, FIG. 4. Appearance of DPPC SUV sample (left-hand vial), interdigitated sheet sample (center vial), and IF-liposome sample (right-hand vial). The DPPC concentration was 20 mg/ml for each case in NaCl–Tris buffer. The DPPC interdigitated sheets were formed by the addition of 3.0 M ethanol at room temperature. The IF-liposomes were formed by raising the temperature to $55°$ for 15 min. The ethanol was not removed. The DPPC interdigitated sheet sample vial is inverted to demonstrate the extreme viscosity of the sample.

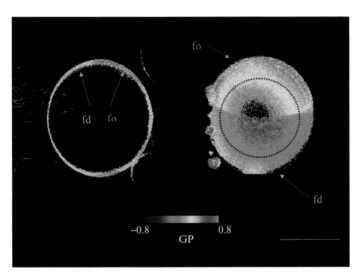

BAGATOLLI, CHAPTER 15, FIG. 6. Two-photon excitation GP images (false color representation) of LAURDAN-labeled GUVs composed DOPC–cholesterol–sphingomyelin (1:1:1, mol/mol) displaying fluid-ordered (fo)/fluid-disordered (fd) phase coexistence (false color representation). GP images obtained at the equatorial region of the GUV, where the photoselection effect is prevented (*left*) and at the polar region of the GUV, where the photoselection operates (*right*). Note the influence of the photoselection effect on the LAURDAN calculated values (dotted black circle) in the image obtained at the polar region of the GUV. Scale bar: 20 nm.